Concentrated Solar Power Systems

IEEE Press
445 Hoes Lane
Piscataway, NJ 08854

Concentrated Solar Power Systems

Bellamkonda Pragathi
DVR & Dr HS MIC College of Technology, Kanchikacherla, India

D. P. Kothari
S.B. Jain Institute of Technology, Management and Research, Nagpur, India

IEEE Press
WILEY

Published by John Wiley & Sons, Inc., Hoboken, New Jersey. Published simultaneously in Canada.

Library of Congress Cataloging-in-Publication Data

Names: Pragathi, Bellamkonda, author. | Kothari, D. P. (Dwarkadas Pralhaddas), 1944- author.
Title: Concentrated solar power systems / Bellamkonda Pragathi, D. P. Kothari.
Description: Hoboken, New Jersey : Wiley-IEEE Press, [2025] | Includes index.
Identifiers: LCCN 2024048156 (print) | LCCN 2024048157 (ebook) | ISBN 9781394272358 (cloth) | ISBN 9781394272372 (adobe pdf) | ISBN 9781394272365 (epub)
Subjects: LCSH: Solar energy. | Solar concentrators.
Classification: LCC TJ810 .P865 2025 (print) | LCC TJ810 (ebook) | DDC 621.47/2 – dc23/eng/20250110
LC record available at https://lccn.loc.gov/2024048156
LC ebook record available at https://lccn.loc.gov/2024048157

Cover Design: Wiley
Cover Image: © PBNJ Productions/Getty Images

Set in 9.5/12.5pt STIXTwoText by Straive, Chennai, India

To my dear friend, Karra Akshitha, whose unwavering support, friendship, and encouragement have been a constant source of strength throughout this journey. Your belief in me has meant the world, and I am forever grateful for your presence in my life.

To my loving husband, S. Hanumantha Rao, whose love, patience, and endless encouragement have been my foundation. Your unwavering belief in me and your support at every step have been instrumental in making this dream a reality. I could not have accomplished this without you by my side.

This work is as much yours as it is mine.

Contents

About the Authors

Dr. Bellamkonda Pragathi is a prominent researcher and educator with a deep expertise in renewable energy systems, machine learning, and Internet of Things (IoT)-based applications. She has contributed extensively to the field through her research on energy management systems, hybrid renewable energy systems, and IoT-driven smart grids. Dr. Pragathi's academic work is widely recognized, and she has co-authored numerous research papers and book chapters on the integration of advanced technologies in energy systems, with a focus on enhancing efficiency, stability, and sustainability.

Dr. Pragathi is also actively involved in mentoring young researchers and has been associated with various international conferences and workshops, contributing to the dissemination of knowledge in the field of renewable energy and emerging technologies. Her collaborative work on CSP systems, alongside Prof. Kothari, represents her commitment to finding innovative solutions for global energy challenges and advancing the frontiers of clean energy technologies.

Together, Prof. D. P. Kothari and Dr. B. Pragathi bring a wealth of knowledge and experience to the study of Concentrated Solar Power systems, offering readers a comprehensive understanding of the principles, challenges, and future directions of this vital renewable energy technology.

D. P. Kothari is a highly respected figure in the field of electrical power systems and renewable energy. With a distinguished career spanning several decades, he has made significant contributions as a researcher, academic, and author. Professor Kothari has held several prestigious positions, including Director General of VITS, Indore, and Director of IIT Delhi, among others. His extensive research in the areas of power systems, energy management, and renewable energy integration has earned him numerous accolades and recognition both nationally and internationally.

Professor Kothari has authored over 50 books and more than 850 research papers in national and international journals, covering a wide array of topics in electrical engineering and renewable energy. His work has inspired generations of engineers and researchers, and his commitment to advancing sustainable energy solutions is reflected in his numerous contributions to the field of renewable energy. Through this work on *Concentrated Solar Power Systems*, he continues to demonstrate his passion for driving innovations in clean energy technologies and addressing the challenges of global energy demand.

Preface

Concentrated Solar Power (CSP) systems represent a pivotal advancement in renewable energy technology, offering a promising pathway for harnessing the sun's vast potential to generate electricity. As the global community increasingly turns toward sustainable energy sources to mitigate the adverse effects of climate change, CSP systems have emerged as a crucial component in the portfolio of clean energy solutions. Unlike photovoltaic (PV) systems that directly convert sunlight into electricity, CSP systems utilize mirrors or lenses to concentrate sunlight onto a small area to generate heat, which is then used to produce electricity. This unique approach to solar energy generation not only provides opportunities for large-scale power production but also integrates thermal energy storage, enabling CSP plants to supply electricity even when the sun is not shining.

The importance of CSP systems lies in their ability to complement other renewable energy sources such as wind and PV by providing dispatchable power. By storing thermal energy, CSP plants can generate electricity during peak demand periods, contributing to grid stability and reliability. This ability to provide on-demand energy addresses one of the main challenges associated with renewable energy—the intermittency of power supply. Furthermore, CSP systems can be integrated with other power generation technologies, such as natural gas or biomass, to create hybrid systems, further enhancing their flexibility and reliability.

This preface introduces readers to the fundamental concepts, importance, and emerging trends in CSP systems. The transition to renewable energy is driven by the need to reduce greenhouse gas emissions, promote energy independence, and create sustainable economic growth. CSP technologies offer a versatile and scalable solution to meet these objectives, particularly in regions with high solar insolation, such as deserts and arid climates. Moreover, advances in CSP technology are continually improving efficiency and reducing costs, making it a more viable option for large-scale energy production.

In this work, we explore the intricate workings of CSP systems, their various configurations, and their applications in today's energy landscape. From parabolic troughs to solar power towers, this text delves into the engineering principles, economic considerations, and environmental impacts of CSP technology. It also addresses the challenges faced by CSP systems, such as the need for significant land area and water resources, while highlighting ongoing research aimed at overcoming these barriers.

As CSP systems continue to evolve, they are positioned to play a vital role in the global transition to a sustainable energy future. This book aims to provide a comprehensive understanding of CSP technology, from its foundational principles to its potential to revolutionize the energy sector. We hope this exploration will inspire further innovation and development in the field, fostering a future where renewable energy is both abundant and accessible to all.

December, 2024

Dr. Bellamkonda Pragathi
Kazipet, Telangana, India

Dr. D. P Kothari
Nagpur, Maharashtra, India

Acknowledgments

We express our deep gratitude to all those who contributed to the completion of this work on *Concentrated Solar Power (CSP) Systems*. The journey to create this comprehensive text has been a collaborative effort, and we are grateful to the many individuals and institutions who provided their support, guidance, and encouragement along the way.

First and foremost, we extend our heartfelt thanks to our families for their unwavering support and understanding. Their patience and encouragement have been the foundation upon which we have been able to devote our time and energy to this project.

We also express our sincere appreciation to our mentors, colleagues, and peers. Their insightful feedback, expert knowledge, and continuous encouragement were instrumental in refining the scope and direction of our work. Your contributions have been invaluable, and this work is a reflection of the intellectual and collaborative spirit we share.

A special thanks to the academic institutions and research organizations that provided the resources and tools necessary for the successful completion of this project. Access to their extensive libraries, databases, and technical expertise has been essential in shaping the depth and accuracy of our research.

We also acknowledge the pioneering researchers and innovators in the field of renewable energy, whose groundbreaking work on CSP systems inspired our exploration of this fascinating and crucial technology. Their commitment to sustainable energy solutions continues to motivate us in our own contributions to the field.

Lastly, we extend our gratitude to the editorial team, the publishers, and everyone involved in bringing this manuscript to fruition. Your attention to detail and professionalism ensured that the quality of our work met the highest standards.

To all those who offered their support, guidance, and encouragement, both directly and indirectly, we thank you. This work is the result of collective efforts, and we are proud to have had the opportunity to collaborate and contribute to the ever-evolving field of renewable energy.

1

Conventional Energy Sources

An important factor to consider when deciding whether a country is considered developed is the amount of energy consumed per person. Maximum energy output is defined by a country's needs, and these needs can only be satisfied if everyone has access to sufficient amounts of energy for things like electricity, transportation, and agriculture. The majority of the world's resources, including coal, oil, and natural gas, are now produced via conventional methods to satisfy global demand. India presently imports about 75% of its crude oil, but this ratio is expected to climb significantly shortly due to the country's expanding economy and rapid expansion. Although only 6% of the world's primary energy is consumed there, India is home to 18% of the world's population. The majority of the world's electricity is generated by burning coal, oil, and natural gas. To create energy, these fossil fuels are burned, which increases the amount of carbon dioxide in the atmosphere and releases several hazardous compounds. The rise in global temperatures is a result of the potent greenhouse gas (GHG) carbon dioxide. The second drawback of using fossil fuels as a source of energy is that their proven reserves have a short shelf life—less than a century, in certain cases. Finding a different electricity production source is required as a result. Power production planning is significantly influenced by two important factors. The principal energy source, which might be any of the three fossil fuels, is an essential component. It is well recognized that all three of these fuels contribute to global warming and environmental harm. The second group of topics relates to the economics of electrical power. Significant losses are experienced during power production, transmission, and distribution. A challenging task that could cut expenses as well as losses in reducing losses or discovering an alternative strategy, like a smart grid or microgrid. The general public and those who create and administer regulations both want access to electricity while avoiding contributing to global warming [1].

Concentrated Solar Power Systems, First Edition. Bellamkonda Pragathi and D. P. Kothari.

1.1 Energy Resources and Their Potential

1.1.1 Oil

Depending on how much oil is used, the world has been split into three categories. Some nations are oil rich, while others are highly industrialized with productive farmland. The majority of the Middle Eastern or Arabian countries are the first group's representatives. The Organization for Economic Cooperation and Development (OECD) is the representative of the other group. There is a third group, which lacks oil and is not as developed as the OECD group. The question of whether oil-producing nations will continue to supply the world with oil until the supply runs out, decide to hoard the oil for their use, or limit sales to nations who support their policies has long been a concern. The Arab nations' lack of food resources led to the current system, in which oil is supplied to industrialized nations. They received food in exchange for the oil, though it wasn't a formal barter deal. The Arab nations gradually became more powerful economically, established new ventures, and began to voice their opinions on global affairs. There is now a kind of unspoken divide between the Arabs and the developed nations. The current terrorism situation and American operations in the Middle East are further highlighting the contrasts, and when the oil taps run dry, the distinctions will become clearer. In the beginning, Arabs were content to receive food in exchange for oil, which helped keep the scales balanced. As can be seen, the exchange rate between food and oil was one bushel of food for every barrel of crude oil between 1950 and 1973, even though the scales had tipped in favor of oil suppliers. Two bushels per barrel in 1974, 5 bushels per barrel between 1975 and 1998, 6 bushels in 1999, and currently 10 bushels or more per barrel of crude oil, which grew by currency rates. Oil is undoubtedly necessary for agriculture and the transportation of food, but the Arab nations have located food sources outside of the OECD [2].

1.1.2 Natural Gas

Over the past 15 years, technology for producing power from natural gas has advanced. By 2020, it's anticipated that natural gas utilization in the creation of power would rise by 87%. According to the most accurate projections, natural gas will support 30% of energy generation in industrialized countries while providing 17% of electricity in emerging nations. Many nations intend to use natural gas since it has been technically developed for use in combined-cycle gas turbines, which are used to generate power. Additionally, while producing the same amount of energy as coal and oil, gas emits less carbon dioxide. It is a cleaner fuel and is utilized in public transportation, such as in Delhi, specifically to preserve the environment. With roughly 33% of global consumption occurring in the former Soviet Union, gas usage for power is already high in those nations. Up to 63% of electricity in the former Soviet Union's nations is anticipated to be produced from gas by 2025. In 2001, the East European nations generated 9% of their electricity from gas; to reach their goal of 50% electricity generation, they plan to import more gas from Russia. Gas consumption for electricity generation in Western Europe increased to 413 billion cubic meters in 2000 and is anticipated to reach 670 billion cubic meters by 2025, having a declining share. Due to the region's use of nuclear energy, the gas portion of power generation is expected to rise from 17% in 2001 to 38% in 2025. Since the 1973 oil crisis, when Western European countries first experienced fluctuating natural gas use. The European Union began limiting the use of gas for generating electricity in 1975. It was 5% in 1981 and stayed there throughout the 1980s. The region began importing gas from Russia, North Africa, and recently discovered sources in the North Sea in the early 1990s; as a result, the share of gas in the electricity market rose, and this pattern is still present today.

Africa and Asia do not yet consume a lot of gas. Nearly one-fourth of the gas consumed by Asian nations is consumed in Japan, which imports all of its gas as liquefied natural gas (LNG). India will end up being a major gas user, and 12.6% of its power is produced using gas.

The majority of the gas used in the United States is imported from Canada, but some gas is also obtained by pipelines from the Alaskan North Sea. From 18% in 2001 to 24% in 2025, gas is anticipated to have a larger part of the electricity market. All of the recently built power plants in the United States, totaling 141,000 MW, are gas based. Overseas imports are probably going to keep rising. In 2002, the United States brought in 4.8 million metric tons of gas or 4% of global consumption. At roughly 46 million metric tons in 2010, the import had more than doubled since 2002. As a result, in addition to importing oil, the United States imports enormous amounts of gas. By 2025, Canada wants to grow its gas-powered electricity generation from 3% to 11%.

The price of natural gas is rising. Prices have more than doubled since 1993. Gas was priced at $2.55 per million British thermal unit (BTU) in 2000, but by 2003 it had risen to $6.31. Russia, Iran, and Qatar, three nations with respective gas reserves of 31%, 15%, and 9%, stand to gain from international gas trading [3].

1.1.3 Coal

The earliest source of energy for producing electricity is coal. It is disliked in developed countries because it creates more carbon dioxide than oil and gas, along with numerous other air pollutants. In addition to carbon dioxide, it also releases sulfur dioxide, nitrous oxide, mercury, and particulate matter. Electricity production uses 64% of all coal production. In 2001, the production was 94.5 Exa Joule (EJ) equivalent. By 2025, it is anticipated that production will rise to 138 EJ. However, all projections indicate that coal will have a decreasing role in the production of power. Although its share of power will drop to 31% by 2025, it was responsible for 34% of electricity in 2001 and 40% of electricity in 2005. By 2025, coal-fired energy will only account for 12% of all electricity produced in Europe, down from 20% now. In 2001, the United States used 40% of the world's coal, compared to 27% combined use by China and India. Coal was used to generate 72% of the electricity in China and India. By 2025, China will produce 73% more power from coal than it does today. However, India's use of coal for energy production will decrease to 63%. By 2025, it is expected that the proportion of coal in US electricity will mostly remain unchanged at 50%.

Coal made up 27% of the electricity in Eastern Europe and the former Soviet Union in 2001. The availability of Russian gas, however, will cause this percentage to decline by 6% by 2025. Eastern Europe is becoming the most polluted region in the world due to the burning of coal. In general, Europe is phasing out the use of coal for electricity. The majority of those coal mines received government funding. Only three nations—the United Kingdom, Germany, and Spain—continue to manufacture hard coal as a result of the European Union's strategy of lowering or eliminating such subsidies. Table 1.1 lists the coal deposits found worldwide.

Every day, a typical 600 MW coal-fired power station will consume 8,193 tons of coal, which may need to be transported over a distance of up to 1,000 km from coal mines. Such a massive amount of coal will need to be transported by regular daily train movements, and there are numerous such factories located all over the world. Although the transportation itself will consume a lot of fuel and produce a lot of carbon dioxide and other pollutants, this calls for a rail system that is effective and has good haulage capacity. 17,900 tons of nitrous oxide, 8,200

Table 1.1 Coal deposits in the world.

Region	Deposit (Gt)	Percent
Asia Pacific	292.5	23.7
North America	257.8	26.2
Former Soviet Union	230	22.4
Europe	125.4	12.7
Africa	55.4	5.6
South & Central America	21.8	2.2
Middle East	1.7	0.2
World	**984.6**	**100**

tons of sulfur dioxide, 4,207,200 tons of carbon dioxide, and 70 kg of mercury are normally produced when coal is burned. Additionally, radioactive substances and fly ash are released by the plants. The factory's modern appearance makes it appear as though it and its surroundings are clean, yet the gases and ash that escape the chimney have an impact on places that are 50–100 km away from the plant. An individual living within 1 km of a coal-fired power station receives a radiation dose that is 1–5% above the typical natural background level, according to studies carried out in the United States between 1975 and 1985. In contrast, residing close to a nuclear power station often results in an increase of 0.3–1% above background levels. Coal-fired power stations' radiation is not controlled in any way. The health impact of mercury is significantly more serious and has not yet been addressed. Mercury is a hazardous, enduring, bioaccumulative contaminant that has an impact on human neurological growth. The effects of airborne mercury are greatest in developing children and infants. It will be challenging to extract mercury from flue gas at a concentration of 1 ppb (part per billion) because the technology for capturing mercury from flue gas has not yet been developed [4].

1.1.4 Hydropower

150 years ago, the Industrial Revolution was powered by the free kinetic energy of flowing water. Water that was flowing and water that was falling from a great height both encountered bladed wheels. The machinery was originally powered by the motion of the water turbines, and then the electrical generators. Because the water turbines could be produced in such huge sizes and the technology for erecting dams was established, which supplied a source of water year-round, hydropower was later totally dedicated to electricity generation. Dams are presently present in several nations, including the United States, China, Brazil, Venezuela, Sudan, Pakistan, and India. Tata built a hydroelectricity plant in India's Western Ghats around the beginning of the twentieth century. When the Bhakra-Nangal hydro project was finished after independence, a huge step was made. After that, several dams were constructed. India has the capacity to produce roughly 84,000 MW of electricity from hydropower. Sources used up until this point total 15,000 MW. Increased hydropower development has mostly as a result of opposition from the general populace. The population is displaced and many species' habitats are destroyed as a result of the dams, which result in a sizable body of water. In addition, silt deposits cause the depth of the water reservoir behind the dam to keep declining. Dredging operations are challenging and expensive to remove the silt. The preference is for small hydro plants over huge ones. Few kilowatt of power are produced by the small plants, which can be enough for a limited area. The best locations for small hydro plants are rivers that flow through steep terrain, such as those in the Indian Himalayan regions.

2,726 TWh were consumed globally in 2000. Despite being the top consumers of hydroelectricity—together using 43% of it—Canada, Brazil, the United States, and China produced roughly 99% of it, and Brazil used it for 87% of its needs. 65% of the hydroelectricity produced worldwide is consumed by industrialized nations. The usage of hydroelectricity is expected to reach 4,000 TWh/year by 2020, with developing countries contributing 800 TWh/year and developed nations providing 600 TWh/year. By 2020, the gap in per capita consumption will widen while the disparity in hydroelectricity generation between industrialized and developing countries will close. Recent years have seen extensive hydroelectricity generation facility construction in Central Asia, South America-Caribbean, Sub-Saharan Africa, and South Asia. It is concerning to note that despite the installation of 5,738 TWh/year hydro capacity in 1997, the amount of power produced from this source was only 2,600 TWh/year, representing a capacity loss of 55% [5].

1.1.5 Nuclear Energy

Contrary to popular assumption, nuclear energy offers a cleaner kind of energy. It does not cause the greenhouse effect and does not release airborne particles. Nuclear fuel requires significantly less of it to create the same quantity of energy than any fossil fuel, and its delivery generates much less air pollution. A nuclear power plant costs less than a power plant using renewable energy sources (RES), but it is still more expensive than any other type of thermal power plant. The issue with nuclear power, however, is that it is not widely accepted because of the widespread belief that these plants are a source of lethal nuclear radiation and that the risk will continue to exist due to nuclear waste from spent fuel.

Although it was the United States that encouraged nations like Japan and France to rely on nuclear electricity, the public perception has generally been influenced by the United States' policy of abandoning the nuclear pathway for electricity in light of widespread public sentiment against nuclear power. Then, in the collective unconscious, there is a mental association between nuclear energy and the atomic bomb, which decimated the two cities of Japan and tragically brought an end to World War II. The concept that terrorists could obtain nuclear weapons and use them against any group of people has mainly been propagated to the American public. As a result, they are opposed to using nuclear energy altogether.

The Japanese were keen not to fall behind in this technology as the world was on the verge of a post-World War II economic and industrial rebirth. The research and technology organization in Japan was granted a $1.88 billion budget to begin with. In 1970, the first industrial power plant started operating. Currently, Japan's 41 commercial reactors provide 33,000 MW of electrical power or roughly 27% of the nation's power needs. The United Kingdom sold Japan its initial gas-cooled, graphite-moderated nuclear reactor. The Japanese were the first to import pressurized water reactor (PWR) and boiling water reactor (BWR) designs when Westinghouse and General Electric of the United States offered them. Then, under license, General Electric and Westinghouse began producing them. Although it was first done for show, Japan now sees nuclear energy as essential because it imports 80% of its energy in the form of Chinese coal and Middle Eastern oil. In terms of global electricity consumption per person, Japan ranks fourth.

The French government may have felt the need to create its Atomic Energy Commission (CEA) as soon as the war ended in 1945 due to a lack of energy supplies. Unfortunately, they focused more on developing bombs than nuclear electricity, returning to it only after the test nuclear device explosion in 1960. At that time, three gas-graphite commercial nuclear reactors and one PWR were also being built. The public, bureaucrats, and politicians all put up a lot of resistance against Electricité de France (EDF) and CEA. After French President Giscard d'Estaing proclaimed that nuclear electricity was the government's policy in 1977, the way toward nuclear energy became apparent. In the late 1970s, several plants were established. In France, nuclear energy produces 75% of the nation's electricity. France has 54 reactors that generate 63,000 MW of power or 4,931 kWh of electricity per person. With 2,265 kWh/person in the United States and 1,450 kWh/person in Japan, this is a highly favorable comparison. France has created reprocessing facilities for old nuclear fuel that are used by numerous European nations as well as Japan. Finland, Norway, and Germany have made the decision to reduce their nuclear electricity production. Germany is of the opinion that it can rely on renewable energy. A German politician said that if Germany couldn't create all the energy it required, it would import it from France. Under the direction of Prime Minister Jawaharlal Nehru and Homi Bhabha, India's CEA was created in 1948, marking the country's entry into this field. The commission, which is currently known as the Bhabha Atomic Research Centre (BARC), has conducted research and development in the field of nuclear science and engineering. In 1969, Tarapur's nuclear power station, built by General Electric, began to run. In 14 plants, it currently generates 2,700 MW of electrical power. By 2020, this is anticipated to rise to 20,000 MW. The issue is that India only has very limited access to the primary uranium-containing mineral, despite the possibility of using thorium, which is more frequently available. Success will be based on how well thorium is used as a nuclear fuel. Meanwhile, India must rely on alternative fuel sources, and agreement with the United States has turned into a political problem. Nevertheless, it is preferable for the government to support a nuclear program for producing electricity. Universities and laboratories should encourage research. India's preference for nuclear power has the dual benefits of generating revenue and preventing the combustion of fossil fuels, which is the main contributor to the greenhouse effect and climate change. We could maintain our environment cleaner by not producing carbon dioxide, sulfur dioxide, nitrous oxide, and particulate matter.

India now generates 1,650 MW of power from its 17 reactors, which have a combined capacity of 4,120 MW. Due to an unstable and insufficient fuel supply, these reactors are not being operated to their maximum potential. India continues to rely on coal for electricity, which is quite harmful in terms of GHG, air pollutants, and particle matter. 55% of Indian electricity comes from coal, and if RES choices like solar, biomass, wind, and hydro and nuclear electricity do not become available, that percentage will rise in the future. Even if sufficient fuel supplies are guaranteed, meticulous calculations by the BARC limit nuclear electricity to 10,000 MW, despite highly optimistic forecasts that position the contribution from the nuclear source at 40,000 MW by 2020. Even without assistance from any other nation, the administration intends to add electricity-generating facilities with greater capacity. Continuing examples include the work of two units each of 220 and 1,000 MW located at Rawatbhata, Kaiga, and Koondankulam, respectively. Although the economy of scale may encourage larger size units, it will be best to keep the unit size small, approximately 220 MW will be better to prevent severe power shortages in case of failure. India's business community has begun making plans to switch to nuclear power. Negotiations between Reliance Power and top nuclear power companies abroad have already begun. The business is looking to join forces in strategic alliances. After the nuclear electricity business is opened to private investment, Tata Power, another energy behemoth in the private sector, aims to either go alone or in a joint venture. Tata Power intends to make a Rs. 12,000 crore initial investment. A major producer of power equipment in the nation, Bharat Heavy Electricals (BHEL), intends to spend Rs. 1 lakh crores over the next five years to enhance its nuclear power infrastructure. The manufacturing behemoth Larsen and Toubro (L&T) also intends to manufacture nuclear equipment. A total of 40 businesses, including Videocon and the Sajjan Jindal Group, are in negotiations to work together on a 2 lakh crore rupee investment. Delegations from Nuclear Supplier Group (NSG) nations have begun to visit India to investigate the market for nuclear electricity. The NSG waiver has raised stock prices for many companies and generally cheered up the stock market. All of these ambitions, however, are subject to the successful conclusion of an Indo-American nuclear agreement. Figure 1.1 depicts India's power generation across the nation, while Figure 1.2 provides a visual representation of the country's installed capacity [6].

Figure 1.1 Regional electricity board of India with their installed generation capacity.

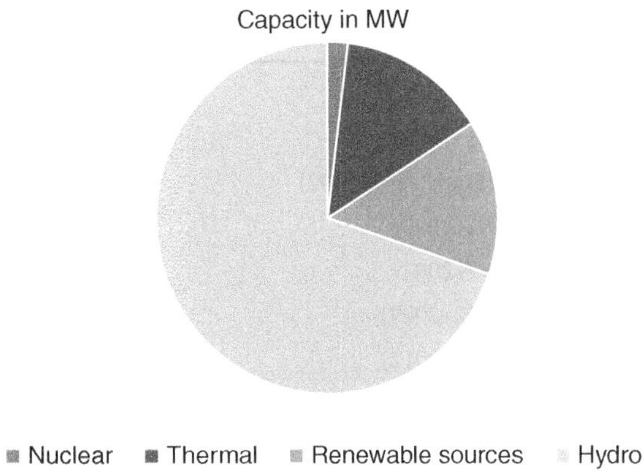

Figure 1.2 Pictorial representation of India's installed capacity.

1.2 Need for Renewable Energy Sources

The development of a country's economy, society, and industry depends largely on its access to energy. Energy consumption is anticipated to increase significantly over the next few years compared to what it is now. India is a rapidly developing nation with significant industrial expansion, but there is also a severe lack of power in the nation. Up until 2000, India produced the majority of its electricity from conventional sources; however, after learning the negative effects of producing electricity from these conventional sources, India began producing electricity from RES. Numerous investments have been made over the last 10 years, and as a result, RES now accounts for more than 14% of all installed power in India.

Recent years have seen a significant expansion in the importance of RES as a resurising energy demand with minimal environmental impact. RES are renewable, efficient, secure, and sustainable sources of energy. The current gap between supply and demand may be filled by RES while also supplying tidy and clean energy. India has a significant RES potential. Power generation from solar, wind, and bio has high potential. Table 1.2 displays the percentage of RES and the total installed capacity of the Indian grid for various five-year plans.

The expansion of RES during several five-year plans is depicted in Figure 1.3. Growth from RES increased significantly during the 11th and 12th plans compared to earlier plans. Figure 1.4 shows the increase in RES during various plans as well as the growth of the total installed capacity. With considerable installed grid-quality renewable power reaching 45 GW as of June 2016, the renewable

Table 1.2 Share of renewable energy sources (RES) and total installed capacity of Indian grid at different plans.

Plan period	Share of RES (MW)	Total installed capacity (MW)
6th Plan (1980–1985)	0	42,584.72
7th Plan (1985–1990)	13.14	63,636.34
2nd Annual Plan (1990–1992)	31.88	69,065.19
8th Plan (1992–1997)	902.01	85,797.37
9th Plan (1997–2002)	1623.39	103,43.04
10th Plan (2002–2007)	7760.6	132,323.21
11th Plan (2007–2012)	24,502.45	199,877.03
12th Plan (2012–April 30, 2016)	43,086.82	302,832.2

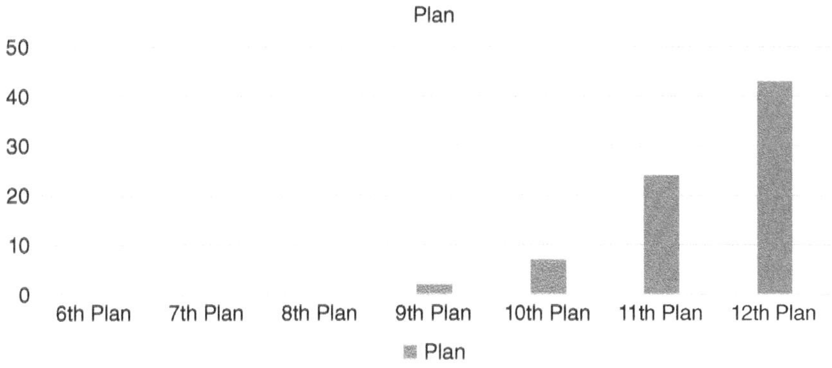

Figure 1.3 Growth of renewable energy sources (RES) during different plans.

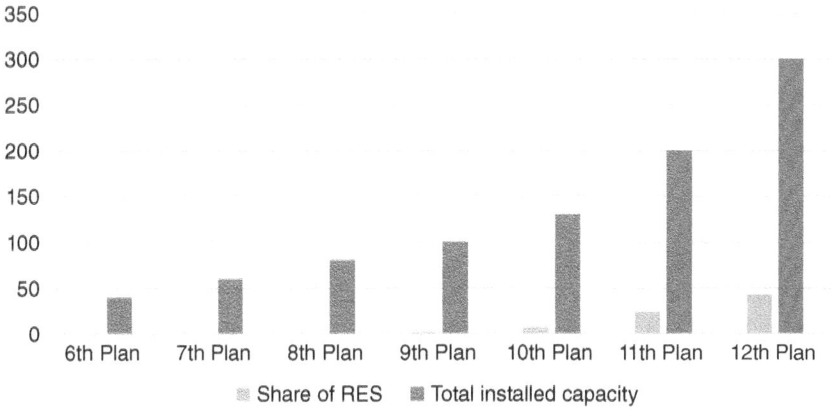

Figure 1.4 Growth of total installed capacity and growth of renewable energy sources (RES) during different plans.

energy sector is now a well-known one in India. Numerous decentralized renewable energy systems, including solar water heating systems, biomass gasifiers, solar photovoltaic (SPV) systems, and others, have been supported by the Ministry of New and Renewable Energy (MNRE) through various programs. To bring electricity to rural parts of India, the Indian government has already launched several programs. In India, SPV is a free resource that is widely accessible, and it is being developed for use in several government programs, including the Jawaharlal Nehru National Solar Mission (JNNSM), state government policies, the National Solar Mission, and others. With these programs, the Indian government hopes to offer electricity to all citizens as well as create power with lower environmental carbon emissions [7].

1.3 Potential Renewable Energy Sources (RES) for Power Generation

1.3.1 Solar Energy

India has 300 clear days each year, which is said to be extremely favorable for the production of solar energy. In the majority of the country, India receives solar energy on average in the range of 5–7 kWh/m^2. Realizing the enormous potential of solar, wind, and biofuels, the Indian government setup a distinct ministry called the Ministry of New and Renewable Energy (MNRE), which has launched a number of initiatives to lower the cost of using these resources to generate electricity. There have been new plans revealed to use the country's enormous solar energy potential, and incentives have been offered to build solar power plants. The prime minister officially announced the Jawaharlal Nehru National Solar Mission on January 11, 2010. To lower the cost of solar power generation in the nation through (i) long-term policy, (ii) large-scale deployment goals, (iii) aggressive research and development (R&D), and (iv) domestic production of crucial raw materials, components, and products to achieve grid tariff parity by 2022, the mission has set the lofty goal of deploying 20,000 MW of grid-connected solar power by that year. India will become a solar energy leader on the planet thanks to the mission's creation of an accommodating governmental environment. It has been announced that a plan will be put into action to create solar parks and ultra-mega solar power projects across the nation starting in 2014–2015 and continuing through 2018–2013. The plan aims to give solar energy generation a major boost by serving as a model project demonstration facility to inspire investors and project developers. This will encourage additional projects of a similar nature and spur economies of scale for cost savings, technological advancements, and significant decreases in GHG emissions. States would be able to satisfy their solar renewable purchase obligation (RPO) deadline, attract major investment from project developers, and give the local populace job possibilities. By preventing emissions equal to the installed capacity and generation of the solar park, the state will also lessen its carbon impact. Additionally, the state will refrain from purchasing costly fossil fuels to run conventional power facilities [6, 8–11].

1.3.2 Wind Energy

India has made significant progress when it comes to producing electricity from wind. India is ranked fourth in the world, having 27 GW of installed wind capacity as of June 2016. The government is promoting the building of wind-based power plants in the nation and supporting these plants by creating a welcoming atmosphere. The southwest summer monsoon, which begins in May or June and draws cold, humid air into the land, and the lesser northeast winter monsoon, which begins in October and moves cool, dry air toward the ocean, both have an impact on the wind in India. Except for the eastern shore, the wind is consistently strong along the whole Indian coast from March to August. From November through March, wind speeds are generally low, yet around the coast of Tamil Nadu, stronger winds are occasionally seen. The generation potential in Tamil Nadu is significantly larger than in other states thanks to these two monsoon winds. On a macro level, the potential for wind power generation for grid interaction has been estimated to be at 45,000 MW based on data gathered from 10 states while only taking 1% of the available land into account.

1.3.3 Biomass Energy

Given its advantages, biomass has traditionally been a significant source of energy for the nation. It is carbon-neutral, renewable, broadly accessible, and has the potential to significantly increase employment in rural regions. Additionally, biomass is able to produce stable energy. More than 70% of the population of the country relies on biomass for its energy needs, and it still accounts for about 32% of the nation's overall primary energy use. The MNRE has recognized the potential and importance of biomass energy in the Indian context and has launched a number of projects to support effective technologies for its use in various economic sectors to guarantee that the greatest possible advantages are obtained. In India, the biomass power production sector attracts annual investments of over Rs. 600 crores, producing more than 5,000 million electrical units and providing more than 10 million man-days of employment each year in rural areas. Under a program for biomass power and cogeneration, bagasse-based cogeneration in sugar mills and biomass power generation have been adopted. The primary goal of this initiative is to promote technologies for the best utilization of the nation's biomass resources for grid power generation. Bagasse, rice husk, straw, cotton stalks, coconut shells, soya husks, deoiled cakes, coffee trash, jute waste, groundnut shells, sawdust, and other biomass products are used to generate electricity. India currently has access to around 500 million metric tons of biomass annually, according to estimates. According to studies funded by the ministry, there is an annual surplus of biomass available to fulfill agriculture and forestry needs of between 120 and 150 million metric tons, or a potential of roughly 18,000 MW. If the 550 sugar mills in the nation adopted technically and economically optimal levels of cogeneration to extract electricity from the bagasse they produce, an additional 7,000 MW of power might be produced through bagasse-based cogeneration. Andhra Pradesh, Tamil Nadu, Karnataka, and Uttar Pradesh are among the states that have assumed leadership roles in the deployment of cogeneration projects. Karnataka, Andhra Pradesh,

Maharashtra, Tamil Nadu, and Chhattisgarh are the states spearheading biomass energy projects.

1.3.4 Hydropower Plants

One of the MNRE's focus areas for electricity generation from renewable sources is the Small Hydro Electricity (SHP) program. It has been acknowledged that small hydropower projects can significantly improve the nation's overall energy situation, especially for distant and inaccessible places. Small hydro projects are being promoted by the government in both the public and private spheres. Grid-interactive and decentralized projects are both receiving equal attention. By the end of the 12th plan, the ministry wants the installed SHP capacity to be around 7,000 MW. The SHP program's main goals are to minimize equipment costs, boost its dependability, and locate projects where they will have the most impact on capacity utilization. The task of creating SHP projects with a station capacity of up to 25 MW has been given to the MNRE. About 20,000 MW of power may theoretically be produced in the nation from such plants. The majority of the potential is in projects based on rivers in the Himalayan states and on irrigation canals in other areas. The SHP initiative is now primarily driven by private financing. The projects are barely economically viable, and the private sector is very interested in funding SHP initiatives. With increased project capacity, these projects become more viable. The ministry wants to capture at least 50% of the nation's potential during the next 10 years.

1.3.5 Hydropower Project Classification

Small and large hydropower projects are the two main categories used to describe hydropower projects. Hydro projects with station capacity of up to 25 MW in India have been classified as SHP projects. The MNRE is given the responsibility for minor hydropower projects (up to 25 MW), while the Ministry of Power is in charge of large hydro projects. Additional classifications for small hydropower projects are shown in Table 1.3.

Table 1.3 Classification of small hydropower plants.

Class station	Capacity in kW
Micro hydro	Up to 100
Mini hydro	101 to 2,000
Small hydro	2,001 to 25,000

1.3.6 Geothermal Energy and Its Potential in India Wave Energy

In many places of the world during the past few decades, geothermal energy has been successfully serving both industrial and home energy needs. It is one of the promising alternative energy sources. Geothermal derives from the Greek terms geo, which means "earth," and therme, which means "heat." So, geothermal energy is simply ground heat. It is a safe and reliable source of power. From the moderate to low temperatures of hot spring systems to the hot rock found a few miles below the earth's surface, and even further down to the extremely high temperatures of molten rocks, geothermal energy resources can be found in various locations. Magma is a layer of hot, molten rocks found beneath the earth's crust. There is constant heat production, primarily from naturally radioactive minerals like uranium and potassium decaying. From the earth's surface, heat is released interior. Normally, the earth's crust protects us from the heat inside. The outer core is liquid, the inner core is solid, and the mantle is semimolten. The amount of heat inside 10,000 m of the earth's surface is 50,000 times more energy than all the world's reserves of oil and natural gas, which is an intriguing point to make here. Since the Roman era, when the heat from the earth was used to heat homes and/or warm bathwater in place of fire, geothermal energy has been one of the most ancient natural sources of heat. Currently, it is primarily employed along plate borders as a source to generate energy. India has a great deal of potential to contribute significantly to the production of affordable, environmentally sustainable geothermal energy. Geothermal energy has the potential to produce about 6.5% of the world's electricity generation, and India will need to step up its efforts in this area in the years to come. However, geothermal energy production is still in its infancy in India. About 340 geothermal hot springs have been identified in India by the Geological Survey. Most of them are in the direct heat application-friendly low surface temperature range of 37–90°C. The Himalayan (Puga, Chhumathang), Saharan (Puga, Chhumathang), Cambay basin, Son-Narmada-Tapi (SONATA) lineament belt, West Coast, Godavari basin, and Mahanadi basin are the seven geothermal provinces in which these springs are distributed. Puga Valley and Chhumathang in Jammu and Kashmir, Manikaran in Himachal Pradesh, Jalgaon in Maharashtra, and Tapovan in Uttarakhand are a few of the well-known geothermal resources. Tattapani, in the state of Chhattisgarh, has also been identified as a new source of geothermal energy. Additionally, Gujarat is planning to use the resources found in Cambay, which is located between the Narmada and Tapi rivers, to generate geothermal electricity. Puga, in the Ladakh region of Jammu and Kashmir, some 180 km from Leh over the vast Himalayan range, is regarded as a promising prospective geothermal energy source. The boiling point of the hot springs in Puga Valley ranges from 30 to 84°C, and they can release up to 300 l of water every minute. In the Puga Valley, a total of 34 boreholes with depths ranging from 23.5 to 384.7 m have been drilled. Hot springs, hot tubs, sulfur condensates, and 4 km-high borax evapotranspiration are all examples of thermal manifestations. The greatest discharge rate from a single thermal spring is 5 l/s, and the hottest thermal spring registers a temperature of 84°C.

1.3.7 Wave Energy

In the form of waves, tidal currents, marine currents, and temperature gradients, the oceans, which make up 70% of the earth's surface, represent a tremendous amount of energy. Our current energy needs are far outstripped by the energy potential of our seas and oceans. India has a lengthy coastline that includes estuaries and gulfs where powerful tides can move turbines for the purpose of generating electricity. The globe is now developing a wide range of diverse methods to capture this energy in all of its forms, including waves (40,000 MW), tides (9000 MW), and thermal gradients (180,000 MW). Although there is now just a small amount of deployment, there is the potential for the industry to expand, spurring economic growth, lowering carbon emissions, and creating jobs not only along the coasts but also inland along its supply chains. We should look into every option we can to encourage innovation, spur economic growth and new jobs, and lower our carbon footprint as the Indian government intensifies its efforts to complete its renewable energy and climate change objectives after 2022. Given the long-term energy demand from this abundant source, action on R&D must be initiated now to ensure that the ocean energy industry can contribute significantly to the accomplishment of our goals in the ensuing decades. The MNRE scans the horizon for an emerging technology that appears promising and examines the range of alternatives accessible to aid in its development. In more than 30 nations, more than 100 distinct ocean energy technologies are now being developed. The majority of innovations are currently in the demonstration or early commercialization stages.

1.3.8 Tidal Energy

The sole type of energy that directly results from the relative motions of the earth, moon, and, to a lesser extent, the earth and solar systems is tidal power. The tides are formed as a result of the Earth's rotation and the tidal forces generated by the moon and sun. Tidal power, sometimes referred to as tidal energy, is a type of hydropower that uses the energy of tides to generate electricity or other useful forms of power. Tidal power has the potential to provide electricity in the future, while not being widely employed now. In comparison to wind and solar energy, tides are more reliable. The availability of sites with high enough tidal ranges or flow velocities has historically been few, which has hampered the use of tidal power as a source of renewable energy overall. However, numerous recent technological advancements and improvements, in both design (such as dynamic tidal

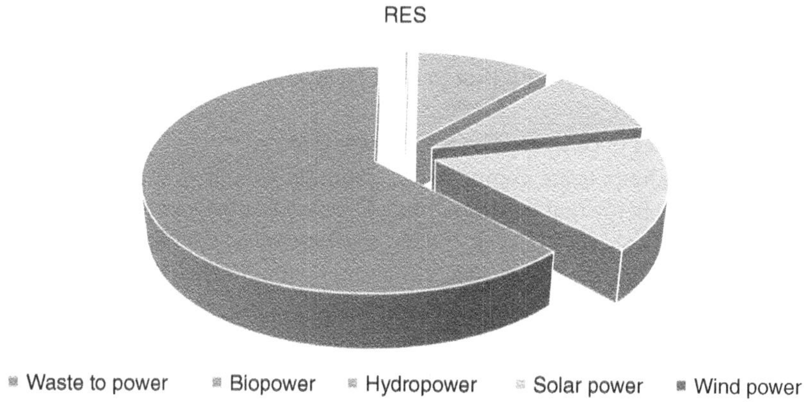

Figure 1.5 Pictorial representation of India's renewable energy sources (RES) installed capacity.

power and tidal lagoons) and turbine technology (such as new axial turbines and crossflow turbines), suggest that the total availability of tidal power may be much higher than previously assumed and that financial and environmental costs may be reduced to competitive levels. Tidal power is essentially limitless and is categorized as a RES since it results from forces brought on by the Earth's rotation, and gravitational interaction with the moon, and the sun. The Rance Tidal Power Plant, constructed at La Rance, France, over the course of six years from 1960 to 1966, was the world's first tidal power facility. Its installed capacity is 240 MW. In Figure 1.5, the installed RES capacity as of 2016 is shown.

- In the Gulf of Kutch or Khambhat, a tidal power plant with a capacity to generate more than 250 MW is anticipated to be built by the British tidal energy business Atlantis Resources.
- The Gulf of Kutch and the Gulf of Cambay in Gujarat and Ganga delta in the Sundarbans, the largest mangrove forest in the world, are the three sites identified as potential areas for tidal power generation.
- A 3-MW plant proposed at the Durgaduani Creek in the Sundarbans delta of West Bengal would be the first attempt by India to harness tidal power for generating electricity.

1.3.9 Off-Grid Renewable Power

Two significant items affect the renewable energy priorities of India as compared to the priorities of other nations. First, renewable energy supplies provide power to those rural populations that are inaccessible through conventional transmission grids, such as hills, islands, and far-flung areas away from a generation center—perhaps the remote areas can only get electricity through renewable sources. Second, RE, being a distributed mode of power generation, is accessible to 1.25 billion to generate their power in their backyards [9].

1.3.9.1 Approaches to Concentrating Solar Power (CSP)

Solar radiation's direct beam component is captured by CSP systems. They are unable to exploit energy that has been diluted by clouds, dust, or other causes, in contrast to flat plate photovoltaics (PV). They are therefore ideally suited to regions with a large proportion of clear sky days, in settings without smog or dust. According to the amount of deployment, the following configurations are currently in use commercially:

- parabolic trough
- central receiver tower
- linear Fresnel
- Paraboloidal dishes
- Fresnel lenses (for CPV)

Every technology boasts a unique set of benefits and, occasionally, niche markets. All varieties of CSP technologies are being actively pursued by project and technology developers. A solar furnace setup is frequently employed as a tool for research initiatives in addition to these ideas that are used commercially. A paraboloidal dish set in a laboratory building with one or more external heliostats aiming sun radiation at it at a fixed angle constitutes the standard component of a solar furnace.

1.4 Concentrating Optics

Solar radiation: At its surface temperature of 5,760 K (5487°C), the sun is a black body radiation source roughly speaking. The sun spectrum has been selectively absorbed at different wavelengths by the different components of the atmosphere by the time it reaches the surface of the planet, and 5,200 K black body behavior is a decent approximation.

Only the portion of solar radiation that is directly emitted is used by concentrating devices; diffuse or scattered radiation is not collected. The flux density of direct (unscattered) sunlight measured on a flat plane perpendicular to the sun's rays is known as direct normal irradiance (DNI). In discussions of solar technology, the terms insolation, radiant flux, flux density, and irradiance are frequently used interchangeably to refer to the rate of solar radiation energy flow through a unit area of space, with SI units of W/m^2 (symbol G). According to Kopp and Lean, the solar constant—the intensity on a plane outside the earth's atmosphere—measures at about 1,367 W/m^2. Although the maximal terrestrial DNI varies greatly depending on the area and the weather, it is frequently accepted to be 1,000 W/m^2.

The process of concentrating solar energy to greater intensities is the brain of a CSP system. The concentration ratio is a crucial measure. There are various ways to define concentration ratio, but two are frequently used.

The optical concentration ratio, C_o, is the proportion of the incident solar irradiance G to the irradiance at the receiver surface, G_r:

$$C_o = \frac{G_r}{G} \tag{1.1}$$

A flux distribution's peak, where the maximum light intensity and concentration ratio are present, is specifically mentioned when defining C_o at any position along the output flux distribution.

The ratio of the collector aperture area to the receiver area A_r is known as the geometric concentration ratio C_g.

$$C_g = \frac{A_c}{A_r} \tag{1.2}$$

The receiver area picked for analysis determines C_g. Receiver apertures can either be constrained to just collect the more intense portion of the focal zone radiation or be selected by design to capture as much of it as possible. Concentration ratio values are sometimes expressed as the number of suns; for instance, a geometric concentration ratio of 1,200 might be referred to as "1,200 suns." This would equate to an average of 1.2 MW/m^2 at the receiver surface at a solar flux of 1,000 W/m^2, as is commonly believed. The sun subtends a half-angle of 4.65 milliradians (mrad) when viewed from the earth. However, the intensity distribution at the sun's borders is not a straightforward step function; therefore, the precise half-angle is relatively complicated.

The circumsolar ratio (CSR) is a crucial metric and is given as

$$\text{CSR} = \frac{G_{cs}}{G_{cs} + G_s} \tag{1.3}$$

where G_{cs} is the solar intensity integrated over the annulus from 4.65 mrad to the outer extent of the solar aureole (surrounding glow), taken as 2.5° (42.6 mrad) for the convenience of easy measurement using laboratory equipment, and G_s is the solar intensity integrated over the solar disk from 4.65 mrad to its limit at 4.65 mrad.

Due to a greater percentage of flux spillage, high CSRs can drastically lower the capture efficiency of high-concentration collectors. The distribution of the sun's form (intensity vs angle) changes along with the CSR. The amount of particulate matter or moisture in the air, in particular, has a significant impact on the actual form of the sun.

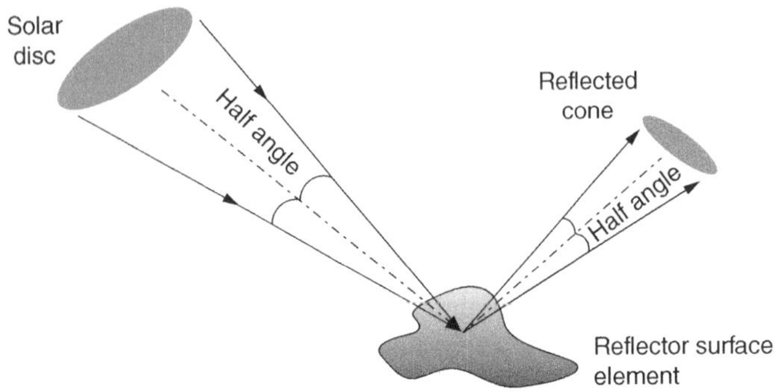

Figure 1.6 Points on a reflector surface reflect direct solar irradiation in a cone of rays.

There have been several different sun-shape distributions proposed; one that is frequently used is that by Buie et al. [12], which provides: a detailed parameterization of the circumsolar radiation distribution around the solar disk. This model incorporates different sun-shape factors to better simulate the actual spread of solar radiation, particularly under varying atmospheric conditions. It is widely applied in solar energy research for accurately modeling solar concentrator performance and assessing energy yield in solar thermal systems and I_r is the solar intensity measured relative to the intensity at 0°, in terms of the sun. The Kuiper sun shape [12] and that of Rabl and Bendt [13] are two other sun-shape distributions that are utilized for modeling. Except for when modeling concentrators have very high optical concentration ratios (10,000 suns), as in dish concentrators or solar furnaces, the precise sun-shape distribution employed for modeling purposes makes little effect, assuming that the optical surface is also very accurate. A flat-topped intensity distribution, often known as a pill-box sun shape, is adequate for working with lower-concentration collectors.

Calculation of sun position: A solar concentrator must be oriented to the direction of the incident sun rays, whether it is line focusing or point focusing as shown in Figure 1.6.

1.5 Limits on Concentration

Smaller receiver apertures and lower thermal losses are advantages of higher concentration levels. But there is a basic difference.

The concentration achievable is constrained by thermodynamics. Specific concentrator geometries can further limit it, and then several practical considerations can further limit it still. Winston et al. [14] officially present the limitations of concentration. An overview of the principles and findings is provided in this section.

1.6 Conclusion

Conventional energy sources, such as coal, oil, natural gas, and nuclear power, have been the backbone of global energy supply for decades. They have provided reliable and large-scale energy production, powering industries, homes, and transportation systems. However, the reliance on these energy sources has come at a significant environmental cost, contributing to GHG emissions, air pollution, and resource depletion. Additionally, the finite nature of fossil fuels poses long-term sustainability challenges.

As the world faces growing environmental concerns and the pressing need for sustainable energy solutions, it has become increasingly clear that a transition away from conventional energy sources is necessary. While conventional energy will likely remain part of the energy mix in the near future, the focus is now shifting toward integrating cleaner RES such as solar, wind, and hydropower. Advancing energy storage technologies and enhancing energy efficiency will also be critical in reducing dependency on traditional energy sources. Ultimately, the global energy transition aims to create a more sustainable, resilient, and environmentally responsible energy future.

References

1 Cheng, F., Zhang, Y., Zhang, G. et al. (2024). Eliminating environmental impact of coal mining wastes and coal processing by-products by high temperature oxy-fuel CFB combustion for clean power generation: a review. *Fuel* 373: 132341.

2 Esiri, A.E., Kwakye, J.M., Ekechukwu, D.E. et al. (2024). Leveraging regional resources to address regional energy challenges in the transition to a low-carbon future. *Open Access Research Journal of Multidisciplinary Studies* 8 (1): 105–114.

3 Liu, A., Imran, M., Nassani, A.A. et al. (2024). Reducing carbon emissions with Geoscience solutions: a look at the contributions of nuclear energy, technology, and Green Finance. *Geoscience Frontiers* 15 (4): 101698.

4 Cano-Londono, N.A., Medina, O.E., Mozo, I. et al. (2024). Viability of the steam-based extraction of extra-heavy crude oil using nanoparticles: exergy and life-cycle assessment. *Energy* 304: 131929.

5 Farhana, K., Mahamude, A.S.F., and Kadirgama, K. (2024). Comparing hydrogen fuel cost of production from various sources-a competitive analysis. *Energy Conversion and Management* 302: 118088.

6 Ibekwe, K.I., Etukudoh, E.A., Nwokediegwu, Z.Q.S. et al. (2024). Energy security in the global context: a comprehensive review of geopolitical dynamics and policies. *Engineering Science & Technology Journal* 5 (1): 152–168.

7 Chu, H., Huang, Z., Zhang, Z. et al. (2024). Integration of carbon emission reduction policies and technologies: research progress on carbon capture, utilization and storage technologies. *Separation and Purification Technology* 343: 127153.

8 Francisco López, A., Lago Rodríguez, T., Faraji Abdolmaleki, S. et al. (2024). From biogas to biomethane: an in-depth review of upgrading technologies that enhance sustainability and reduce greenhouse gas emissions. *Applied Sciences* 14 (6): 2342.

9 Mendrela, P., Stanek, W., and Simla, T. (2024). Sustainability assessment of hydrogen production based on nuclear energy. *International Journal of Hydrogen Energy* 49: 729–744.

10 Zhao, Y., Hu, B., and Wang, S. (2024). Prediction of brent crude oil price based on LSTM model under the background of low-carbon transition. *arXiv preprint arXiv:2409.12376* .

11 Onwuka, O.U. and Adu, A. (2024). Eco-efficient well planning: engineering solutions for reduced environmental impact in hydrocarbon extraction. *International Journal of Scholarly Research in Multidisciplinary Studies* 4 (01): 033–043.

12 Buie, D., Monger, A.G., and Dey, C.J. (2003). Sunshape distributions for terrestrial solarsimulations. *Solar Energy* 74: 113–122.

13 Rabl, A. and Bendt, P. (1982). Effect of circumsolar radiation on performance offocusing collectors. *Journal of Solar Energy Engineering* 104 (3): 237.

14 Winston, R., Welford, W.T., Miñano, J.C. et al. (2005). Nonimaging Optics. Elsevier Academic Press, Maryland Heights, MO.

2

Measurement and Estimation of Solar Irradiance

2.1 Introduction

Solar irradiance, the power per unit area received from the sun in the form of electromagnetic radiation, is a critical parameter in the design, operation, and efficiency of solar energy systems. Accurate measurement and estimation of solar irradiance are essential for assessing the potential of solar energy in a given location, optimizing the performance of photovoltaic (PV) systems, and ensuring reliable energy production. Solar irradiance data is also valuable for climate studies, agricultural planning, and the assessment of environmental impacts.

There are two primary components of solar irradiance: direct solar irradiance, which measures the sunlight that reaches the earth's surface without being scattered, and diffuse solar irradiance, which accounts for sunlight scattered by atmospheric particles and clouds. The sum of these components forms the global solar irradiance. Understanding both direct and diffuse irradiance is crucial for the optimal placement and orientation of solar panels to maximize energy output.

Traditional methods of measuring solar irradiance involve the use of pyranometers and pyrheliometers, which provide high-precision readings. However, due to the spatial and temporal variability of solar irradiance, it is often impractical to rely solely on direct measurements. Therefore, various estimation techniques have been developed, ranging from empirical models based on historical data to advanced computational methods, such as machine learning algorithms, satellite-based observations, and weather forecasting models. These techniques enable more comprehensive and accurate predictions of solar irradiance over larger areas and longer time periods, thus supporting the planning and management of solar energy projects.

As the global demand for renewable energy sources grows, improving the accuracy of solar irradiance measurement and estimation has become a focus of both academic research and industry practice. This introduction sets the stage for discussing various techniques and tools used in the measurement and estimation

Concentrated Solar Power Systems, First Edition. Bellamkonda Pragathi and D. P. Kothari.

of solar irradiance, highlighting the importance of accuracy in maximizing the efficiency of solar energy systems [1, 2].

2.2 Parabolas and Paraboloids

The parabola's capacity to concentrate parallel radiation to a point f distances from its vertex gives it a major role in solar concentrators. The cross-sectional shape of a linear concentrator in the form of a trough is provided by the parabola, a two-dimensional curve. Similar to the paraboloid, a parabolic surface of revolution, dish concentrators have this surface form.

In Figure 2.1, the functional relationship that characterizes a parabola with its axis parallel to the y-axis is

$$y = \frac{x^2}{4f} \tag{2.1}$$

while it is true for a paraboloid whose axis coincides with the z-axis

$$z = \frac{r^2}{4f} = \frac{x^2 + y^2}{4f} \tag{2.2}$$

The axis and a line drawn from the focus to the actual edge of the concentrator are at an angle known as the rim angle ($\emptyset_R$). A parabolic concentrator's cross-sectional geometry is totally determined by its focal length and rim angle together. A parabola's or a paraboloid's rim angle is determined by

$$\tan\emptyset_R = \frac{W/2}{f - z_R} = \frac{4fW/2}{4f^2 - (W/2)^2} \tag{2.3}$$

in which W denotes the parabola's width and z_R its rim depth.

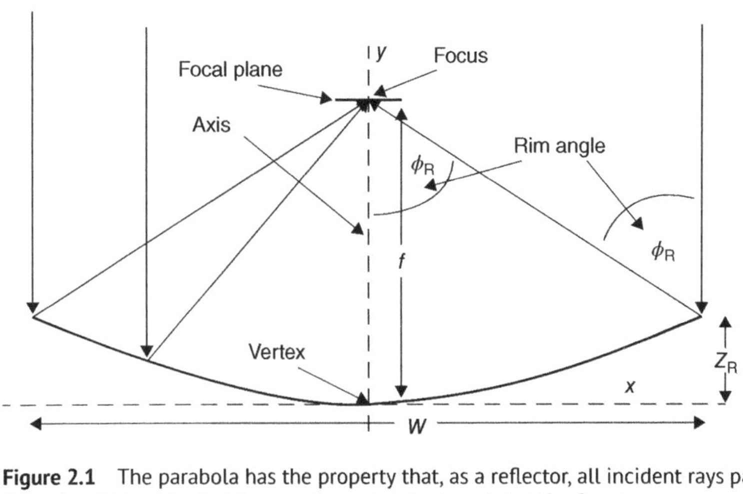

Figure 2.1 The parabola has the property that, as a reflector, all incident rays parallel to the axis will be reflected to pass through a single point at the focus.

Only perfectly parallel incoming radiation causes the parabolic effect, which concentrates light into a single point. Due to the physical size of the solar source, sunlight has various incident angles, as was already mentioned. As detailed in the following sections, this lowers the optical concentration that can be achieved at the focus of parabolas and paraboloids.

Limits with flat receivers: Each point on a parabolic mirror will reflect a cone of rays with half-angle sizes that correspond to the angular distribution of the solar source. When incident on a flat target positioned in the focal plane, consider the size of the spot created by the cone of rays formed by the points on the mirror (Figure 2.2). The largest such spot will be created by the rim's rays.

The reflection point's x distance from the axis is

$$x = 2r\sin\emptyset_{\mathrm{R}} \tag{2.4}$$

and because of reflection from this point, the focal spot's width on the focal plane will be

$$d = \frac{2r\sin\theta_{\mathrm{s}}}{\cos\emptyset_{\mathrm{R}}} \tag{2.5}$$

An elliptical spot with a major axis length of d will really be formed on the target by the reflected cone from a single spot on the mirror. Both dishes and troughs fall within this definition.

The total effect of all the spots across the mirror will not accurately represent the image of the sun created at the focal plane since it is obvious that the size of the focal-plane spot relies on the received incidence angle. The term

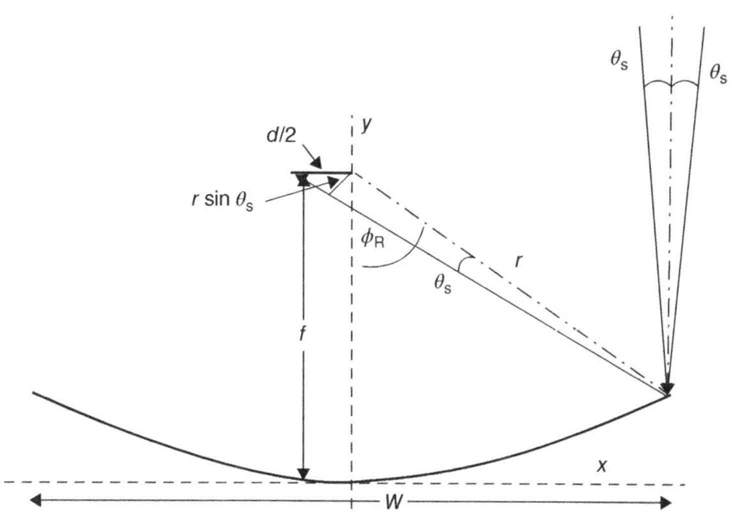

Figure 2.2 Concentrating solar radiation with a perfect parabolic mirror to a flat target.

"nonimaging optics" refers to solar concentrator optics for this reason. It will roughly generate a correct picture of the sun in the focal plane if the rim angle $\varnothing_R$ is small, at which point r is about equal to f over the entire mirror, and the image diameter will be

$$d = \frac{2f\sin\theta_s}{\cos 0} = 2f\sin\theta_s \tag{2.6}$$

The mirror elements at the very edge of a dish or trough with a flat receiver and a 90° rim angle, on the other hand, will create an indefinitely long spot that spreads radiation throughout the whole focal plane.

The diameter of the receiver will be determined by the size of the reflected spot from points at the edge of the mirror with $x = \mathrm{W}$ and $\varnothing = \varnothing_R$ if the receiver is large enough to accept reflected spots from the entire mirror surface. A parabolic trough with a flat receiver will consequently have a geometric concentration ratio of

$$C_g = \frac{A_C}{A_R} = \frac{LW}{Ld} = \frac{W}{d} \tag{2.7}$$

This results from the substitution of Eqs. (2.5) and (2.6).

$$C_g = \frac{\sin 2\varnothing_R}{2\sin\theta_s} \tag{2.8}$$

Take the derivative with respect to $\varnothing_R$ to determine the ideal rim angle for a parabolic trough

$$\frac{dC_g}{d\varnothing_R} = \frac{d}{d\varnothing_R}\left(\frac{\sin 2\varnothing_R}{2\sin\theta_s}\right) = \frac{2\cos 2\varnothing_R}{2\sin\theta_s} = 0 \text{ at } \varnothing_R = 0, 45^0, 90^0, \ldots \tag{2.9}$$

The highest concentration ratio corresponds to $\varnothing_R = 45°$ with a solar acceptance angle of $\theta_s = 4.65$ mrad and offers a maximum concentration ratio 1:1 for a trough with a flat receiver.

$$C_{g,\text{trough}} = \frac{1}{2\sin\theta_s} \approx 108 \tag{2.10}$$

On applying this study on a paraboloidal dish,

$$C_g = \frac{A_c}{A_R} = \frac{\frac{\pi}{4}W^2}{\frac{\pi}{4}d^2} = \left(\frac{W}{d}\right)^2 = \left(\frac{\sin 2\varnothing}{2\sin\theta_s}\right)^2 \tag{2.11}$$

From which,

$$\frac{dC_g}{d\varnothing} = \frac{d}{d\varnothing}\left(\frac{\sin^2 2\varnothing}{4\sin^2\theta_s}\right) = \frac{\sin 2\varnothing\cos 2\varnothing}{\sin^2\theta_s} = 0 \text{ and gives maximum value at } \varnothing = 45^{\circ} \tag{2.12}$$

Hence

$$C_{g,dish} = \frac{1}{4\sin^2\theta_s} \approx 11{,}600 \tag{2.13}$$

These values fall under the thermodynamic constraints by around half (trough) and a quarter (dish). The rim angle is used by the majority of dish and trough concentrators with cavity receivers, which have flat opening close to 45°.

Limits for cylindrical and spherical receivers: Utilizing a receiver with a circular cross section, as shown in Figure 2.3, is an additional option to think about. In this scenario, the target's diameter must

$$d = 2r\sin\theta_s \tag{2.14}$$

$$C_g = \frac{A_c}{A_R} = \frac{LW}{L2\pi r\sin\theta_s} = \frac{\sin\emptyset}{\pi\sin\theta_s} \tag{2.15}$$

The ideal trough rim angle, determined by solving for the maximum geometric concentration ratio as previously, is $\emptyset = 90^0$, and this angle results in

$$C_{g,trough} = \frac{1}{\pi\sin\theta_s} \approx 63.5 \tag{2.16}$$

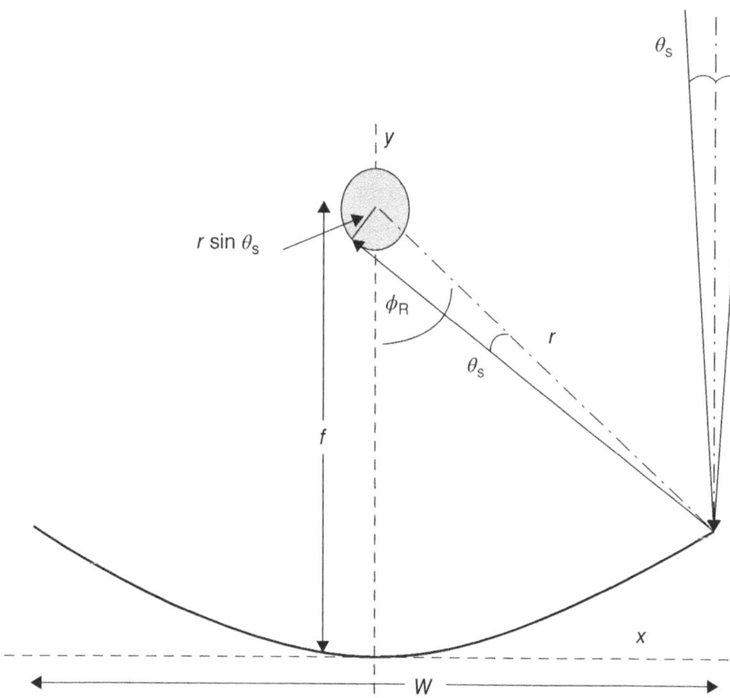

Figure 2.3 Concentrating solar radiation with a perfect parabolic mirror to a circular cross-section target.

Rim angles that are close to $\emptyset = 90^0$ are consequently used in trough concentrators with cylindrical evacuated tube receivers. Similarly, for a dish with a spherical receiver,

$$C_g = \frac{A_c}{A_R} = \frac{\frac{\pi}{4}W^2}{4\pi r^2 \sin^2\theta_s} = \frac{\frac{\pi}{4}4r^2\sin^2\emptyset}{4\pi r^2 \sin^2\theta_s} = \frac{\sin^2\emptyset}{4\sin^2\theta_s} \tag{2.17}$$

and once more, the maximum geometric concentration ratio happens at a $\emptyset = 90^0$ angle.

$$C_g = \frac{1}{4\sin^2\theta_s} \approx 11{,}600 \tag{2.18}$$

The limit determined by the optical analysis is then equal to the thermodynamic limit divided by for troughs and equal to 1/4 of the thermodynamic limit for dishes.

The derivation above shows that the local contributions to focal spot size vary as a function of reflection radius x, so the geometric concentration ratio limits are lower than the optical concentration ratio limits at the focal spot's center. Take note that this analysis is of geometric concentration ratio.

Additional findings by Rabl include the mirror area per aperture area for these four various collector setups. Since flat receiver troughs and dishes reflect light at a more normal angle on average than those with cylindrical or spherical receivers, they need fewer mirrors for a given aperture.

Secondary optics: According to Gordon, the addition of secondary optics, a second stage of optical concentration, can produce concentration ratios that are up to 80–90% of the thermodynamic limit (Figure 2.4). The one-stage concert is held at, because of growing surface errors due to the considerable distance between the mirror and receiver and increasing cosine losses per glass area, smaller angles of the order of 80° that are frequently employed in practice.

A secondary Trombe–Meinel cusp concentrator, shown here in the Novatec linear Fresnel collector design, is used to enhance the concentration of sunlight

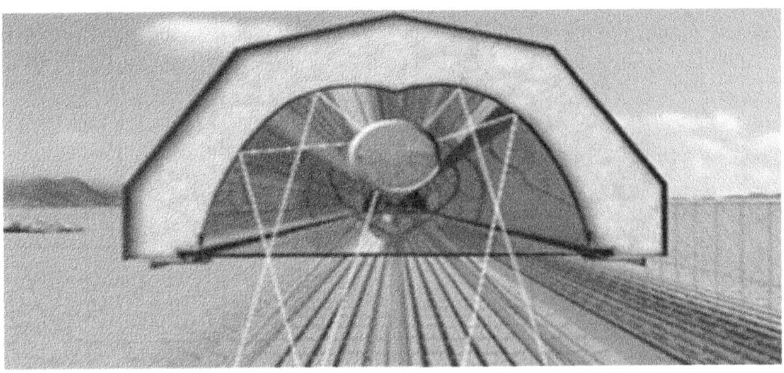

Figure 2.4 A secondary Trombe–Meinel cusp concentrator.

onto the receiver. This type of concentrator employs a reflective cusp geometry to capture and direct additional solar radiation that would otherwise miss the primary absorber, thereby improving the optical efficiency and overall energy yield of the system. The incident radiation to the receiver of the trators described above frequently only comes from a rather small angular range (for a flat absorber, a half-angle of 45°). Therefore, conservation of étendue implies that greater concentration is feasible; in reality, this is accomplished by utilizing various funnel-like mirror systems. Geometric arrangements can be utilized to achieve this additional level of concentration, such as the compound parabolic concentrator (CPC, also known as a Winston collector), the Trombe–Meinel cusp, and the Mouchot conical mirror.

2.2.1 Practical Factors Reducing Concentration

The amount of concentration that can be attained in solar concentrators is lowered by various error-causing factors. Starting at the scale of microns and working up to the scale of centimeters, they can be divided into groups based on the size of the flaw generating the inaccuracy.

2.2.1.1 Specularity Error

Specularity is a reflector's ability to reflect light like a mirror, or more specifically, how closely reflected rays adhere to the law of reflection, which states that the reflected angle must equal the incident angle. A diffuse reflector, on the other hand, disperses reflected light in numerous directions. Such surfaces enable a photograph to be taken to record the irradiance in the focal plane of a dish, heliostat, etc., making them valuable in approaches for measuring concentrator performance. The specularity error σ_{sp}, which results from the microscopic characteristics of surfaces, is present to some extent on all real surfaces. It is defined locally on a surface as the standard deviation of the distribution of reflected ray angles at a given incidence angle.

2.2.1.2 Surface Slope Error

Local ripples and distortions will appear on concentrator mirrors at a slightly larger scale. The degree of aberration is frequently referred to as the surface slope error, which is defined as the standard deviation of the angular deviations of the surface normal vectors from their ideal directions, sampled across the surface of the mirror.

The optical performance of the concentrator suffers with an increase in focus spot spread and a fall in maximum concentration ratio when surface slope error increases. Although values up to around 5 mrad may be sufficient, depending on the type of concentrator and application, values of about 0.4 mrad are typically observed in very accurate systems.

2.2.1.3 Shape Error

On a wider scale, facets are typically used to build concentrators. These facets may be orientated improperly, or the overall structure may be distorted as a result of thermal expansion, wind loads, or the release of residual stresses. These mistakes can be precisely measured and translated into an effective shape error, or sh, which is the standard deviation in the surface normal angles resulting from these various types of misalignment. Accurate measuring techniques, such as photogrammetry, are used to do this.

2.2.1.4 Tracking Error

The concentrator should always be correctly pointed at the sun via a tracking system. In practice, tracking systems are imperfect and can fail to direct the collector precisely toward the sun. With an "on–off"-type tracking control system in particular, this angular offset frequently changes over time. The tracking error σ_{tr}, defined as the standard deviation of the angular error distribution, can be used to describe the angular error when it is studied over a longer period than the source of the problem.

2.2.1.5 Combinations of Errors

The specularity, slope, shape, and tracking errors of a concentrator can be integrated into its total optical error, or tot, in a typical system. The total optical error can be calculated as the root sum of the squares of the component errors if all the sources can be reasonably accurately described as a normal distribution of randomly distributed errors. This is,

$$\sigma_{total} = \sqrt{\sigma_{sp}^2 + \sigma_{surf}^2 + \sigma_{sh}^2 + \sigma_{tr}^2} \tag{2.19}$$

This equation should only be used with caution since some forms of surface and shape errors might not be effectively described by a normal distribution. Additionally, the square root of the aforementioned equation can have an additional term 2 inserted inside it if the sun shape is modeled as a simple normal distribution.

2.2.1.6 Cosine Losses and End Losses

The fact that mirrors can't always be perfectly oriented to the incident sun rays is a significant source of "loss" in solar concentrators. The apparent area of a mirror, as seen from the sun, decreases as it reflects off-axis in proportion to the cosine of the incidence angle. This decrease in apparent area, which is referred to as a cosine loss even though technically the energy was never collected in the first place, directly lowers the concentrator's concentration ratio assuming that the aperture area of the concentrator is equal to the mirror area.

All types of solar concentrators have this cosine loss effect. Cosine losses vary depending on where the sun is in relation to the concentrator, with the exception of dishes and lenses. The only cosine losses are caused by the curved mirrors because dishes are always directed directly at the sun. The off-axis placement of most central receiver systems is done thus in an effort to reduce average cosine losses.

Particular to trough and linear Fresnel concentrators are end losses. These are the terms used to describe the radiation that is reflected from the mirrors but misses the receiver since the sun is not directly overhead of the collector and is instead concentrated beyond the end of the receiver. These losses may be present year-round or only in the mornings and evenings, depending on the orientation (east-west or north-south).

Commonly, end losses and cosine losses are taken into account while calculating the optical in η_{optical}.

2.2.1.7 Focal Region Flux Distributions

There will be a spread of irradiance in the focal region once the light has been refracted or reflected along its passage through a concentrator. For receiver system design and performance analysis, the focal region flux distribution must be predicted and measured.

2.2.1.8 Prediction of Focal Region Distributions

Ray tracing is a popular method for predicting the dispersion of the focal zone. In this computational method, the collector geometry is combined with data on the sun's shape, specularity, slope, shape, and tracking faults to create a virtual "scene" that includes every element and its corresponding properties. The virtual "scene" is then "fired" with a large number of simulated rays whose angular distribution fits the designated sun form. The path of every light is visible in the scene. A portion of the ray's strength is attenuated by an obstruction by the surface reflectivity, and the direction in which it is reflected is determined by a random sample taken from the slope error distribution and the specularity distribution. "Tracing" is carried out repeatedly until the ray disappears or is entirely absorbed. In general, it is also possible to accommodate reflections through glass, etc. On several "virtual sensor surfaces" around the scene (which often correspond to a receiver surface or the focus plane), a grid of cells is defined. The strength of all the rays incident on each grid cell of the sensor surface can then be added to determine the radiation distributions. The accuracy of the calculation is improved by employing smaller grid cells and more rays, but the calculation time is prolonged. To analyze a solar concentrator with adequate accuracy, several million rays are often needed.

In the Concentrating Solar Power (CSP) industry, popular general-purpose ray-tracing programs like ASAP, OptiCAD, OSLO, and ZEMAX are used. Tona-tiuh, a free open-source ray-tracer being actively developed at CENER in Spain with a focus on solar concentrators, is a recent competitor. DELSOL, MIR-VAL, UHC, HFLCAL, FIAT LUX, and SolTrace are a few optical codes designed especially for solar applications. Some of these codes employ a convolution approach, which treats the distribution of rays as a whole rather than computing each ray's path individually through the scene; the approach is said to be within 1–2% of detailed ray tracing but is much faster to calculate.

Measurement of focal region distributions: Direct measurements of focal region distributions can be made utilizing various camera-targeting technique. At the concentrator's focal plane is a Lambertian target that has been appropriately chilled. Images of the target's reflected radiation are then taken using a camera. The irradiance at the target can be calculated if the surface's reflectivity is known and the camera is properly calibrated.

Figure 2.5 displays an illustration of a distribution that was experimentally calculated from the ANU 500 m^2 dish system and has a Gaussian-like distribution.

It is possible to calculate the percentage of solar radiation that is first blocked by the concentrator aperture and then blocked by the receiver by superimposing a genuine receiver geometry on a known focus region distribution. The optical efficiency of the system is significantly influenced by this capture fraction or intercept

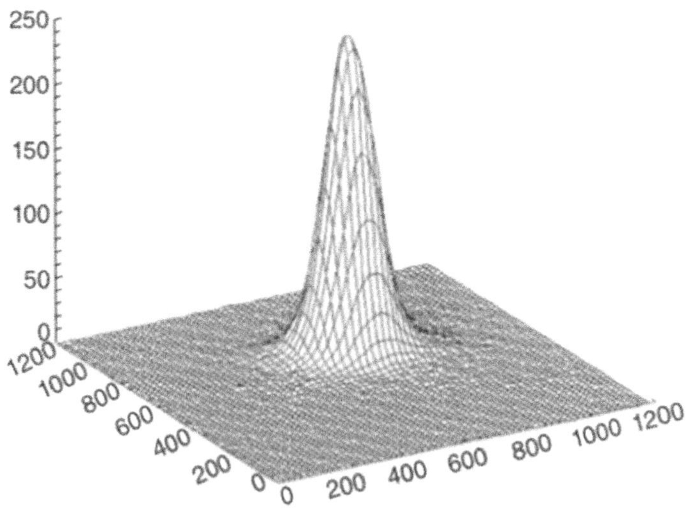

Figure 2.5 Experimentally determined irradiance distribution of the ANU 500 m^2 dish at the focal plane. Spatial units are millimeters, the vertical scale uncalibrated relative units.

factor. The intercept fraction of a circular aperture imposed on the flux pattern in figure is shown as a function of diameter.

The distribution of the focal zone is spread out as a result of optical imperfections, which also lowers the concentration ratio that may be achieved. This necessitates the use of larger receivers in order to prevent flux spillage or the loss of rays that miss the receiver and travel elsewhere. On the other hand, a loss analysis shows that having an excessively large receiver will cause significant radiative and convective losses; hence, for an optimum collector, some degree of flux spillage will be accepted.

The quantity of flux intercepted by the receiver aperture in comparison to the amount intercepted by the collector allows for the determination of the collector efficiency with knowledge of the flux density distribution (Figure 2.6).

$$\eta_{\text{optical}} \equiv \frac{\iint G_{\text{incident}}(t)\mathrm{d}A\,\mathrm{d}t}{\int A_{\text{collector}} G_{\text{sol}}(t)\mathrm{d}t} \tag{2.20}$$

A collector is the aperture area of the collector, and $G_{\text{sol}}(t)$ is the time-varying direct normal irradiation (DNI). $G_{\text{incident}}(t)$ is the time-varying concentrated irradiation at the receiver. The time integrals are no longer required if irradiation is constant during the period in question.

This definition encompasses the following effects: The actual capture fraction of a concentrator that is precisely oriented for a certain receiver, the loss resulting from nonunity mirror reflectivity, the loss from tracking error, and any cosine losses.

2.2.1.9 Losses from Receivers

At the steady state, energy outflows from the flow of heat transfer will balance the net energy flow from focused solar radiation into a receiver.

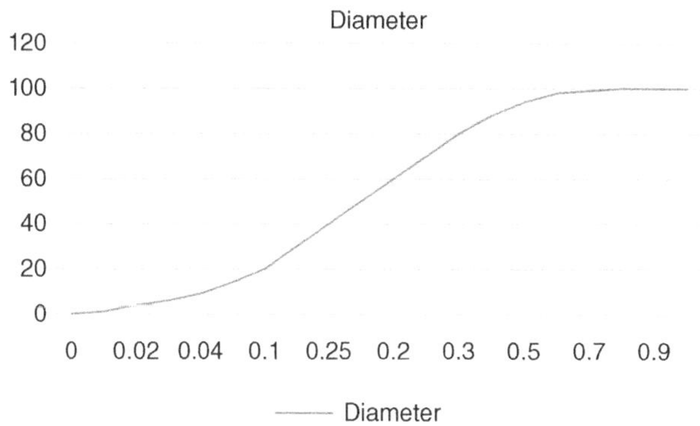

Figure 2.6 Empirical relative intensity distribution of the ANU 500 m^2 dish at the focal plane be assessed at a certain time or averaged over a period of time to detect changes in tracking or structure, for instance.

The aggregate of these four contributions will represent the overall loss.

$$\dot{Q}_{\text{loss}} = \dot{Q}_{\text{ref}} + \dot{Q}_{\text{rad}} + \dot{Q}_{\text{conv}} + \dot{Q}_{\text{cond}} \tag{2.21}$$

The energy efficiency of a receiver is the ratio of the energy that is usefully converted to the input energy from the concentrated that is intercepted

$$\eta_{\text{rec}} = \frac{\dot{Q}_{\text{converted}}}{\dot{Q}_{\text{input}}} = \frac{\dot{Q}_{\text{input}} - \dot{Q}_{\text{loss}}}{\dot{Q}_{\text{input}}} \tag{2.22}$$

where for the receiver

$$\dot{Q}_{\text{input}} = \iint G_{\text{incident}}(t)\text{d}A\ \text{d}t \tag{2.23}$$

The receiver efficiency will be 0 if the beneficial energy conversion process is "turned off," and the receiver will heat up until the total losses perfectly balance the incident radiation. The term "stagnation temperature" is used to describe this temperature.

2.2.1.10 Radiative Losses

The net radiation that is emitted from receivers as a result of their temperature and the reflection of some of the incident concentrated solar radiation are both examples of radiative loss mechanisms. The net energy transfer that occurs is the result of the interaction between the radiation that surfaces emit and absorb. Each surface in a receiver will emit radiation at a rate determined by the black body emissive power multiplied by its emissivity, which is proportional to the fourth power of temperature [3, 4].

Any surfaces it is incident upon will reflect a portion of the incident energy. The geometry will determine how much of this radiation is lost at the receiver. If glass covers are used in the receiver's construction, they will also cause reflection losses that antireflective coatings may be able to reduce. For cavity receivers in particular, radiation reflected or emitted from one portion of a receiver surface is extremely likely to be incident on another, necessitating the simultaneous solution of the entire process in order to determine the net absorbed radiation. This is commonly done via ray-tracing software for incident radiation that is reflected, but not for radiation that is emitted. The techniques for determining the eventual distribution of incident solar radiation on diffusely reflecting surfaces are identical to those needed to figure out how diffuse radiation from hot receiver surfaces is ultimately dispersed [5, 6].

In proportion to the view factor, also known as the radiation shape factor, all other surfaces within the field of vision will partially intercept the radiation that is leaving a surface.

F_{ij} = the fraction of radiation leaving surface i and reaching surface j.

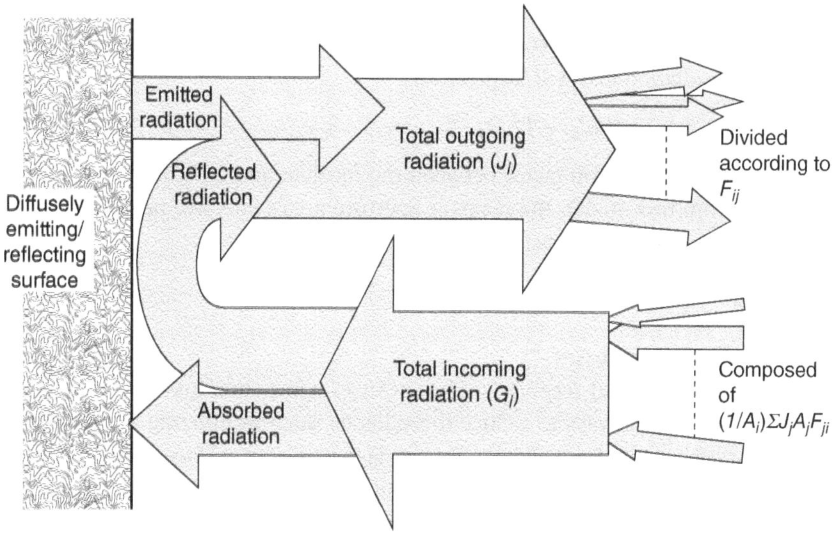

Figure 2.7 Radiation energy balance on a diffusely emitting and reflecting surface.

Calculating the ultimate distribution of absorbed energy becomes a challenge if these other surfaces reflect and absorb different fractions of the incident energy.

Figure 2.7 shows the radiative energy balance of a specific (diffusely reflecting) surface element. G_i represents the total irradiance in this diagram.

A_i is the area of surface i, F_{ij} is the view factor from surface i to surface j, and J_i is the radiosity of surface i, which is defined as the total radiative flux (reflected + emitted) exiting that surface. In this method, each wavelength range (such as solar vs thermal wavelengths) can be treated independently.

A mathematical equation for the energy balance is written for every surface in a radiation exchange network. These equations may be solved easily to determine the net heat transfer at each surface since they combine to produce a linear system in terms of radiosity. Each surface needs boundary conditions; the surface temperature can be fixed and the equation solved for net heat transfer, or it can be done the other way around.

This method can be used to resolve the radiation exchange between surfaces inside a solar concentrator receiver. The first step is to determine from the optical characteristics of the concentrator how much concentrated solar energy enters through the aperture and strikes each surface. To any radiation incident on it from other internal surfaces, the aperture itself can be thought of as a black body surface. Surface temperature and the externally convected flux can be used as boundary conditions for external convective losses [7, 8].

If the aperture is represented as a single surface in a simplified model, interacting just with the surroundings at the average receiver temperature, then the emitted radiation loss will be given by

$$\dot{Q}_{\text{rad}} = \sigma A \varepsilon F_{\text{RS}} \left(T_{\text{rec}}^4 - T_{\text{env}}^4 \right) \tag{2.24}$$

where the simplified form factor between the receiver and environment is called F_{RS}. In a simplified model, the receiver aperture area's effective net absorptivity can be used to approximate reflection losses.

$$\dot{Q}_{\text{ref}} = (1 - \alpha) A \dot{Q}_{\text{sol}} \tag{2.25}$$

2.2.1.11 Convection Losses

Air moving over heated receiver surfaces causes convection losses in solar concentrators. There are ways to reduce these losses, such as utilizing an evacuated glass tube (as in parabolic troughs) or covering the receiver surface with glass. Certain dish and tower systems have quartz windows that can tolerate extremely high solar radiation. In other instances, the receiver is an open chamber, and the only thing acting to reduce convective loss is the buoyancy of the hot air held inside the downward-facing receiver.

Convection losses are challenging to detect and estimate, including forced convection brought on by wind and spontaneous convection brought on by the buoyancy effects of warm air. Convection and radiative loss are related processes, which complicates matters because the magnitudes of both types of losses depend on the surface temperature, which in turn depends on external heat transfer (radiation and convection) and internal heat transfer (to the fluid flowing through the interior of the receiver). Many times, the issue is attempted to be decoupled, for example, by assuming a constant receiver wall temperature or a constant heat flux into the heat transfer fluid (HTF).

Since convection losses are extremely challenging to quantify directly, empirical conclusions are typically obtained by deducting other well-known losses from the overall energy balance. Commercial computational fluid dynamics (CFD) systems like FLUENT and Star-CCM enable in-depth simulations of convection heat transport. Open FOAM is a free open-source substitute (Open FOAM Foundation) to facilitate analysis. Although it is possible to apply semiempirical correlations for natural and forced convection heat transfer coefficients, the accuracy is typically no better than 20%.

The receiver convective loss can be determined once correlations have yielded a value for the average convection heat transfer coefficient.

$$\dot{Q}_{\text{conv}} = hA(T_{\text{rec}} - T_{\text{amb}}) \tag{2.26}$$

2.2.1.12 Conduction Losses

Insulating covers on the back of receivers and any thermal channel between hot receiver surfaces and the surrounding environment will cause thermal losses. These losses are also inversely proportional to an overall "thermal resistance" that varies on material conductivities and shape and is linearly proportional to the temperature differential.

$$\dot{Q}_{\mathrm{cond}} = \frac{(T_{\mathrm{rec}} - T_{\mathrm{env}})}{R_{\mathrm{th}}} \tag{2.27}$$

2.2.1.13 Energy Transport and Storage

To transmit energy, high-temperature HTFs must be moved through piping networks. Energy transport based on chemical reactions is also being developed. HTF networks are significantly smaller for tower systems compared to scattered collector fields like trough systems. We won't go into further detail here because the fundamental methods for calculating pressure drop, pumping power, and heat loss through insulated pipes are well established in engineering practice. It is important to keep in mind, however, that there are conventional methodologies for thermoeconomic optimization of pipe diameter and insulation thickness that should be used, but it is important to keep in mind that what is cost-effective in a CSP system may not be so.

The currently popular commercial method uses two-tank molten-salt systems that incorporate sensible heat transfer with hot and "cold" tanks of molten $NaNO_3$-KNO_2. The majority of other methods that are being developed also require the storage of high-temperature materials. It's crucial to reduce parasitic energy use and direct heat conduction losses. The temperature drop that occurs when thermal energy is sent to storage and then later extracted is a less evident but possibly more significant issue [9, 10].

2.3 Power Cycles for Concentrating Solar Power (CSP) Systems

Various concentrator types can be used with various solar to electric energy conversion devices.

2.3.1 Steam Turbines

Steam turbines are used to generate the majority of the world's power, often using steam created by fossil or nuclear heat sources. The simplicity with which a new source of heat can be incorporated into the preeminent power-generating technology is one of the advantages of CSP. Consequently, steam turbines are used by the great majority of CSP systems that are currently in service. For the purpose of producing steam for use in steam turbine energy conversion, all concentrator types have been used.

A Rankine cycle plant that uses a steam turbine operates as follows: compresses pure feedwater to high pressure (over 10 MPa, for example); boils and superheats steam in a boiler that may be located in the focal point or heated using a heat exchanger with another HTF; expands the steam to low pressure via a series of turbines that power a generator. Like all heat engine cycles, the Rankine cycle has a better conversion efficiency for higher steam temperatures and pressures at turbine input. Including different phases of steam bleed from the turbines, which may subsequently be used to gradually heat feedwater before usage in boilers, is a crucial component that boosts efficiency.

To prevent blade erosion, the proportion of liquid condensing within the turbine must be kept to a low value. By making sure the vapor is suitably superheated before expanding, this can be accomplished. Increasing boiler pressure to boost efficiency may result in materials that cannot be sufficiently superheated to prevent condensation in the turbine. Reheating the vapor after partial expansion solves this issue. A large-scale steam turbine-based power plant often combines all of these qualities, and the overall configuration resembles that in Figure 2.8 in most cases.

Practically speaking, controlling the chemical makeup of the cycle water is a crucial step in the process. To maintain acceptable levels of contaminants, such as dissolved salts, a portion of the water is periodically "blown down" (expelled from the system). Direct mixing of feedwater and bled steam occurs in an open feed-water heater. Prior to supplying feedwater to the boiler, the heating action of operating it at atmospheric pressure also has the effect of driving out dissolved air [11, 12].

Systems that are created as larger units and used at full capacity are more effective. The 50–100 MWe scale is where the majority, but not all, of the size efficiency gain is realized. Larger systems are less expensive per capacity unit. The average size of the large power-generating turbines used in coal power plants is 500 MWe. There is a trade-off between turbine size and CSP application because a larger turbine necessitates a larger field, which leads to longer thermal line losses. Many commentators advise a 250 MWe unit as having the lowest cost of energy.

Modern steam turbines with the highest efficiency operate with a steam input temperature of up to 700°C. Trough and linear Fresnel concentrators, however, are only capable of operating at temperatures of up to 500°C when using direct steam generation (DSG) as opposed to thermal oil as the HTF. Tower and dish systems can achieve the maximum conceivable steam turbine inlet temperatures

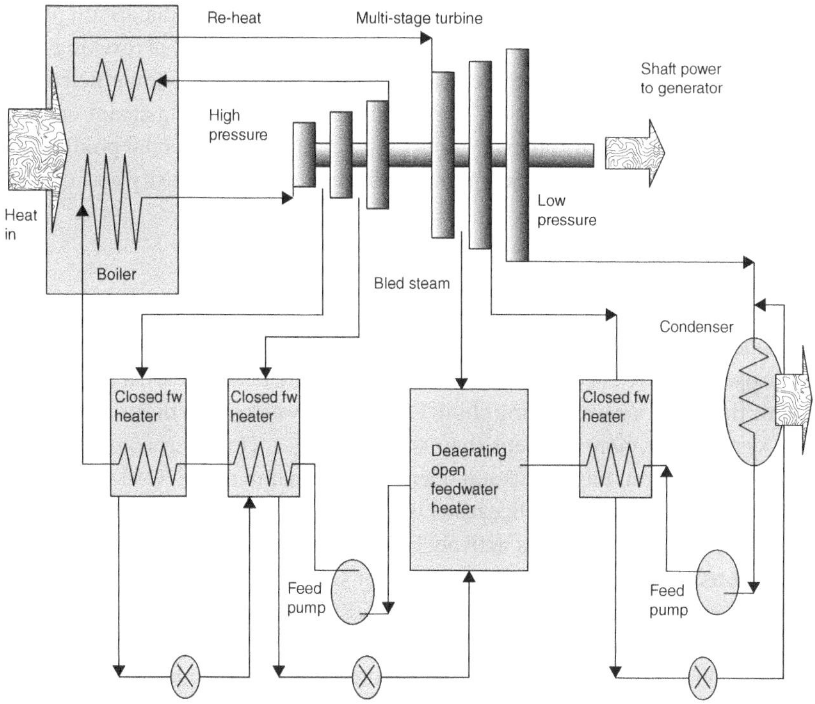

Figure 2.8 Indicative configuration for a large-scale steam turbine power plant.

and pressures; in such case, the restriction is the ability of the materials to survive, either in the turbine or in the solar receiver.

Modern steam turbines are currently made to operate under extremely harsh circumstances for optimal conversion efficiency. The term "supercritical steam" refers to steam that is both pressured and heated above the critical point (22 MPa, 374 °C), where the phase change occurs continuously rather than by nucleate boiling. These turbines have not yet been used with CSP facilities since they are only practical at very large scales.

The intermittent and erratic character of solar input is a key area of distinction between solar and fossil steam turbine operating. This could have two effects: the desire to run at part-load more frequently and the desire to adjust turbine speeds up and down more frequently and quickly than in steady-state fossil fuel operations.

Energy storage can help to some extent to offset these consequences, however, directly implementing techniques and technology from conventional generation may not always produce the best outcomes. With these challenges in mind, turbine manufacturers are now creating steam turbines specifically for CSP applications. These steam turbines may start off cold and attain maximum power in 30 minutes or less, depending on the start temperature. For existing CST facilities, the typical steam turbine heat to AC electricity conversion efficiency is around 40% gross at full load. At very small sizes, reciprocating steam engines are more effective than turbines. Despite their continued commercial production, they are still comparatively uncommon because of their complexity and upkeep problems.

2.3.2 Organic Rankine Cycles

In essence, an organic Rankine cycle (ORC) is identical to a steam Rankine cycle; but, to operate more effectively with lower temperature heat sources, it employs an organic fluid with a lower boiling point. For smaller systems (less than a few MWe), ORC systems are more efficient than steam turbines. However, compared to a water/steam system, the capital and operating and maintenance (O&M) expenses are higher per installed MW. Because ORC technology is better suited to sources with lower temperatures, it is actively being investigated for geothermal power applications. A few small linear concentrator CSP systems have been equipped with ORC systems. Another possible use for ORC systems in CSP is as a "bottoming cycle," in which a high-temperature cycle (such as the Brayton cycle discussed below) generates exhaust heat that would otherwise be wasted but is hot enough to drive an ORC system for additional power generation [13].

2.3.3 Stirling Engines

A Stirling engine is a reciprocating piston engine that operates on a constant, contained volume of gaseous working fluid, often hydrogen or helium but occasionally air. A combination of adiabatic (zero heat transfer) and constant temperature processes make up the perfect cycle. It can achieve the highest thermodynamically feasible convergence efficiencies between two constant temperature limits in the ideal limit. While there are big, fossil-fueled marine propulsion systems, none of the Stirling engines for CST applications that have been developed so far are smaller than tens of kWe. Dish concentrators have historically used Stirling engines. Even though dishes have been used for DSG, PV, and other ways, many in the CSP sector refer to dish systems in general as "dish-Stirling" due to their long history and prevalence. Other collection types have not yet seen any significant application of Stirling engines. There are two different forms of Stirling engines: free-piston engines, which use an oscillating piston connected to a linear generator but without any kind of physical restraint connection, and piston-crankshaft engines constructed similarly to internal combustion engines. Both have been applied in actual life [14].

With total solar to AC electric efficiencies of about 30% at design point DNI, reliably reproducible dish-Stirling systems continue to hold the record for the highest solar to electric conversion efficiency of any technology. Compared to Rankine cycles, Stirling systems can be used for considerably smaller systems; however, thermal storage for the dish-Stirling design has not yet been created.

2.3.4 Brayton Cycles

Jet engines and the turbo generators used in gas turbine power plants are both built on the Brayton cycle. Contrary to popular belief, "gas turbines" are not so named because they burn gas; rather, the name relates to the fact that a gas, typically air, serves as the working fluid. Any hydrocarbon fuel, including jet fuel, diesel, LPG, propane, or biogas, might be used in a fuel-fired mode to produce the necessary heating. As an alternative, solar heat could be employed to boost the compressed air's temperature before expansion. Only tower systems or dishes are anticipated to be able to operate at the temperatures needed for efficient operation, which is roughly 1,000°C before expansion. Brayton cycle demonstration CSP systems that use solar heating have been put to use. A combined cycle power plant uses a gas turbine in fossil fuel applications, and the high-temperature exhaust gases are subsequently diverted to a "heat recovery steam generator" that produces steam for a steam turbine cycle. Potentially exceeding 50%, the combined conversion efficiency is the highest thermal-to-electric conversion efficiency option that is now commercially accessible. The ability to conduct combined cycle operations with either steam or ORC bottoming cycles in a similar high-efficiency way is a key benefit of using the Brayton cycle for CSP applications.

In comparison to Stirling engine systems, the Brayton cycle may result in lower O&M expenses for dish applications. Supercritical carbon dioxide Brayton cycles (s-CO_2 cycles) are one area of current solar thermal energy research interest. Higher overall cycle efficiencies can be attained thanks to CO_2's distinct thermodynamic characteristics compared to air.

2.3.5 Concentrating Photovoltaics

To maximize the return on the investment in concentrator systems, they often use pricey, high-efficiency cells. Single-crystal silicon cells of excellent quality have been used, with efficiencies of about 20%. Beyond this, cells have experienced a sharp rise in efficiency over the past 10 years (from 30% efficient to 43% efficient), such as the multijunction cells created for space applications. There are parasitic losses associated with a CPV system's tracking system, controls, wire losses, inverter efficiency, and cooling system performance. The system's overall useable AC output is decreased by these parasitic losses [15].

Any high-concentration PV system has a major problem in the form of heat produced by photons that are not directly converted into electricity. Without a highly effective cooling system to remove the heat, a triple-junction cell would be destroyed at 500 suns in a matter of seconds.

2.3.6 Others

Other methods of converting solar radiation to energy exist and may one day be more cost-effective, such as:

- The Kalina cycle is a modified Rankine cycle that performs better at temperatures between 200 and 300°C. It uses combinations of ammonia and water with different concentrations. It targets geothermal power uses and is pursued commercially.
- Thermoelectric converters use heat to generate electricity. The operation of semiconductor-based devices is similar to that of PV cells, with the exception that thermal excitation rather than individual photon absorption is used to excite electrons into the conduction band.
- Thermionic converters, which excite electrons from an active surface across an evacuated zone to a collector, also generate energy directly from heat.
- Thermo-photovoltaics convert the radiation released from heated surfaces using PV cells designed for thermal radiation wavelengths.
- Magneto-hydrodynamic converters create a potential difference by expanding hot, ionized gas through a magnetic field.

Maximizing system efficiency: Optical errors, concentration ratios, and the causes of heat losses from a receiver were covered in earlier sections. The aggregate performance of these subsystems yields a system efficiency:

$$\eta_{\text{system}} = \eta_{\text{optical}} \times \eta_{\text{receiver}} \times \eta_{\text{transport}} \times \eta_{\text{storage}} \times \eta_{\text{conversion}} \tag{2.28}$$

The concentrator/optical efficiency is frequently understood to refer to total reflectivity and flux spillage losses up to the point where radiation reaches the receiver. Both reflective losses from the receiver and radiative, convective, and conductive losses are included in the receiver efficiency. The efficiency of transport $\eta_{\text{transport}}$ comprises thermal losses from pipes carrying the HTF to the storage or the power cycle (or the chemical or industrial heating process) as well as losses in heat exchangers between the HTF and the working fluid of the power cycle. Depending on the type of storage, storage efficiency η_{storage} may also take into account pressure or chemical potential losses as well as losses from heat exchangers between the storage medium and the working fluid of the power cycle as well as thermal losses from the storage vessel. The power cycle efficiency (or conversion efficiency $\eta_{\text{conversion}}$), which also takes into account heat, friction, and electrical losses in the power cycle, is finally constrained by the second law of thermodynamics, which is covered in the section below. It should be highlighted that enhancing the efficiency of each component separately will not increase the total system efficiency; the maximum is only discovered by looking at all effects together [8].

2.4 Energy Analysis and the Second Law of Thermodynamics

The ideal outcome would be to maximize the system's overall energy efficiency η. However, analyzing the problems in terms of just energy does not give all the details required for optimization.

The effectiveness of a heat engine's conversion between isothermal reservoirs is ultimately limited by the well-known Carnot limit, which is determined by the second law of thermodynamics, which deals with the intrinsic irreversibility of processes.

$$\eta = \frac{W}{Q_H} = \frac{Q_H - Q_L}{Q_H} = 1 - \frac{Q_L}{Q_H} = 1 - \frac{T_L}{T_H} \tag{2.29}$$

Temperatures in this equation are absolute temperatures with 0 K equaling 272.15°C in SI units of kelvin. Effective power generation cannot be achieved by operating at low temperatures to attain high receiver efficiencies. The only power cycle that has a conversion efficiency as predicted by the Carnot formula is an ideal Stirling engine. A Stirling engine runs between a fixed high-temperature receiver and a set low-temperature environment in a receiver-mounted application at a steady state. The power cycle, however, is no longer functioning between effective isothermal reservoirs in systems where an HTF is used between the receiver and the power cycle. The special situation of the maximum amount of work that may be extracted between two isothermal reservoirs is quantified by the Carnot efficiency. It can be utilized to build the thermodynamic concept of "exergy," which is particularly beneficial.

The highest amount of labor that could be productively accomplished while bringing a system's (thermodynamic) equilibrium with its environment is known as exergy. A system in this context could be as simple as a specific unit mass of a substance, such as an HTF. As a result of this definition, every energy flow has an exergy flow that corresponds to it in terms of quantity. This exergy flow represents the optimum amount of work that could be taken from the energy flow. Exergy analysis is an effective method for comprehending and enhancing CSP systems, which are created with the production of power as their primary output in mind [16].

A matching energetic efficiency can be defined for each subsystem efficiency. Since the current version includes pure exergy as the ultimate output type,

Table 2.1 Thermodynamic efficiency metrics relative to an environment of 298 K, 1 bar.

Temperature (°C)	Temperature (K)	Carnot efficiency (%)	Exergy-energy ratio of a constant specific heat HTF (%)	Exergy-energy ratio of steam at 15 MPa (%)	Exergy-energy ratio of steam at 5 MPa (%)	Typical real steam cycle net efficiency (%)
5,527	5,800	94.9	84.1			n.a.
1,000	1,273	77.0	56.1			n.a.
700	973	63.9	43.3	43.0	44.6	40–45
400	673	56.5	35.9	42.4	40.2	35–40
300	573	43.9	23.8	31.5	31.5	25–30
200	473	33.0	22.0			

maximization of either overall energy or exergy efficiency will take place for the identical design and operational parameter values.

Table 2.1 shows the energy-to-energy ratio of a few hypothetical HTFs at the same temperature together with the temperature dependence of the Carnot efficiency. It is clear that operating thermal systems at temperatures that are really feasible indicates a significant reduction in the ability to extract work when compared to the sun's surface temperature. It can also be seen that the potential to extract work is decreased when heat is transferred from an isothermal source to a real HTF heated to the same temperature. That is, the process of destroying energy is irreversible.

As demonstrated, the Carnot efficiency for operation from an infinite isothermal source is greater than the exergy-to-energy ratio of the alternative working fluids for a given temperature. It is possible to imagine that energy in a substance at a specific temperature is available over the range from that temperature down to ambient temperature [16, 17] based on the quantity required for each degree of temperature increase, effectively defining the thermal energy potential that can be harnessed as the substance cools. This concept helps in understanding and calculating the usable heat energy available within a material, which can be transferred to perform work or drive thermal processes. The transition from water to steam has a significant impact on the profile of steam. Its exergy-to-energy ratio changes in relation to a fluid with constant specific heat, depending on how close this is to the real steam temperature.

The best way to understand the idea of regenerative feedwater heating in a steam cycle using bled steam is from the perspective of exergy. In contrast to letting the highest temperature heat source heat the feedwater from ambient temperature, employing partially expanded steam to preheat the feedwater destroys less energy than doing so. The ability to generate work from the energy absorbed in the receivers is increased when the HTF is sent to the receivers at a higher temperature, which is how we may apply this to a CSP system. However, doing so necessitates maintaining larger HTF flow rates while also increasing the receiver average temperature and associated losses [18].

2.4.1 Heat Exchange Between Fluids

Any location in a system where heat transfer from one fluid to another happens, typically via a heat exchanger, is a significant source of energy loss in CSP systems. Major heat exchange pathways in CSP systems can go from HTF to storage medium to power cycle working fluid to storage medium to power cycle working fluid to ambient air (condenser or cooling tower).

Even though the overall heat transfer coefficient is not high, well-designed heat exchangers produce a broad contact area between two fluid streams, enabling heat transfer to occur with a smaller temperature drop. However, the cost to construct such a heat exchanger can be very high, necessitating a cost-benefit analysis. Systems that eliminate pointless heat exchange processes are becoming more popular, especially in the fields of DSG and molten-salt parabolic trough fields. Even if a heat exchanger achieves a certain overall heat transfer coefficient, the fluids involved could not have the same enthalpy increase to temperature increase characteristics. This is especially true if there is a phase change (for example, from water to steam). This is a significant potential energy loss source. The technique of pinch analysis seeks to optimally combine all heat transfer temperature profiles in complex processes with numerous heating and cooling requirements [5, 19].

Efficiency of a simplified solar collector with 100% optical efficiency, average absorptivity of 0.8 (assumed equal to emissivity) plotted both with no external convective loss (solid lines) and with 20 W/m^2K external convective loss (dotted lines). DNI is 800 W/m^2, ambient temperature 300 K, and sky temperature 270 K.

2.4.2 Optimization of Operating Temperature

To show the competing design requirements, a streamlined study of a solar thermal energy system's performance is helpful. The most efficient power cycles run from high-temperature thermal reservoirs are known from the theory of reversible heat engines, but we also know that thermal losses rise as receiver temperatures rise.

The results for receiver efficiency as a function of concentration ratio and temperature are shown in Figure 2.9 and are obtained by simplifying the formulas for

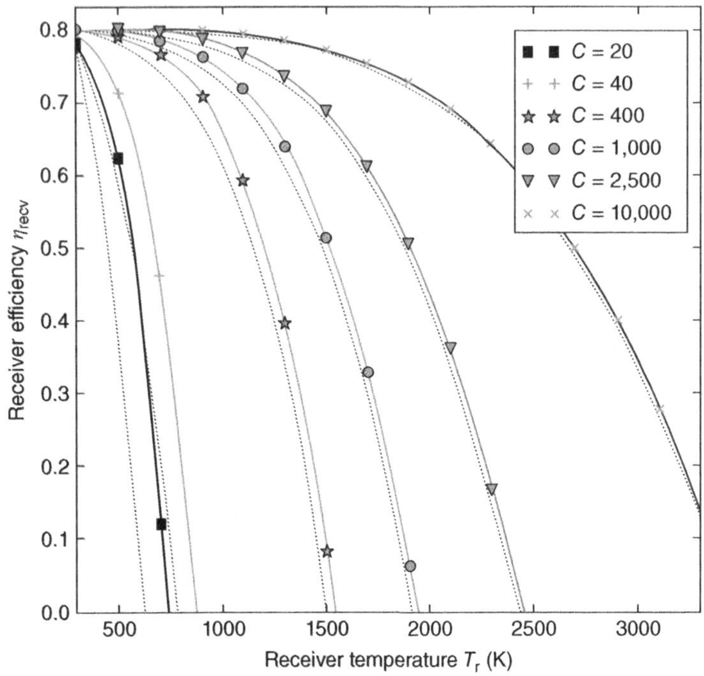

Figure 2.9 Receiver efficiency.

the various loss mechanisms and using some common parameter values. As can be observed, all efficiencies drop to the average absorptivity value at low temperatures. Better concentration systems have better receiver efficiencies at any given temperature, but efficiencies decline as temperature rises. The corresponding stagnation temperature is where the various curves' intersections with the *x*-axis occur.

The receiver efficiency must be increased by a power cycle conversion efficiency in order to provide an overall ideal system efficiency that simulates an optimum operating regime for maximizing power production. Power cycle efficiency's intricate relationship with operational conditions must be determined using empirical performance curves. A good indicator is the difference between the HTF's exergy value before and after heat is transferred to the working fluid at the start of the power cycle, with actual power cycles extracting roughly 70% of the available exergy as work. However, an excellent qualitative representation of the concepts can be found in the Carnot expression [20–22].

The findings are as indicated in Figure 2.10 if the receiver efficiency values of Figure 2.10 are multiplied by the Carnot efficiency relevant to the receiver temperature at each position to obtain a semi-ideal system efficiency. It is obvious

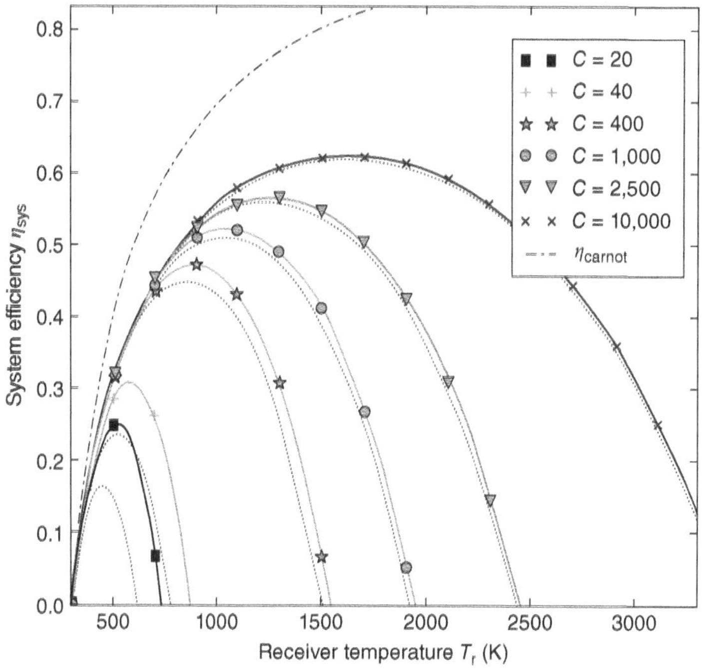

Figure 2.10 Ideal receiver efficiency.

that each concentration ratio now has a distinct peak efficiency at a specific temperature. Peak efficiencies are higher with higher concentration ratios, and they happen at higher receiver temperatures.

Ideal system efficiency as a function of receiver temperature: Carnot efficiency is also shown as a dashed line as a function of receiver temperature. The dotted lines include the effect of convective heat loss, for a fixed convection coefficient as described in Figure 2.10.

2.4.3 Optimization of Aperture Size

The size of the receiver aperture is a crucial parameter to optimize. For a constant receiver temperature, the relationship between radiative and convective losses and aperture area suggests that the aperture area should be kept to a minimum. However, a bigger aperture will block more of the radiation that is incident, boosting optical efficiency. There will be an aperture size beyond which any additional increases will increase thermal losses by more than the collected energy is increased for a given focal-region flux distribution and operational temperature.

Every concentrator and application will require a different selection of ideal aperture and operating temperature. The following findings are from Steinfeld and Schubnell and serve to illustrate the concept. Based on an actual flux distribution observed using a miniature solar furnace, they examined the best options. The real focal-plane flux map is depicted as a contour plot in Figure 2.11. Steinfeld and Schubnell viewed a potential receiver as a uniform temperature black-body disk and simply assumed radiation losses from it. They established an absorption efficiency equal to the sum of concentrator and receiver efficiency based on this premise.

Solar flux distribution was measured at the focus of the Paul Scherer Institute solar furnace in October 1990. The power intercepted by the aperture can be found by integrating solar flux through the circled area (reproduced from Steinfeld and Schubnell).

Energy absorption efficiency as a function of aperture radius for various receiver temperatures: The dotted line is the locus of maximum efficiency, for which optimum aperture radii are determined (reproduced from Steinfeld and Schubnell).

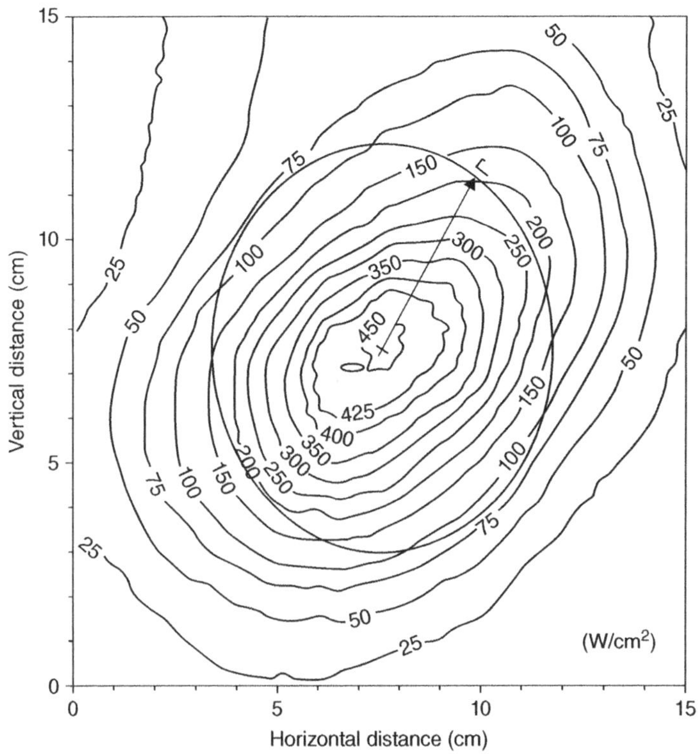

Figure 2.11 Absorption efficiency.

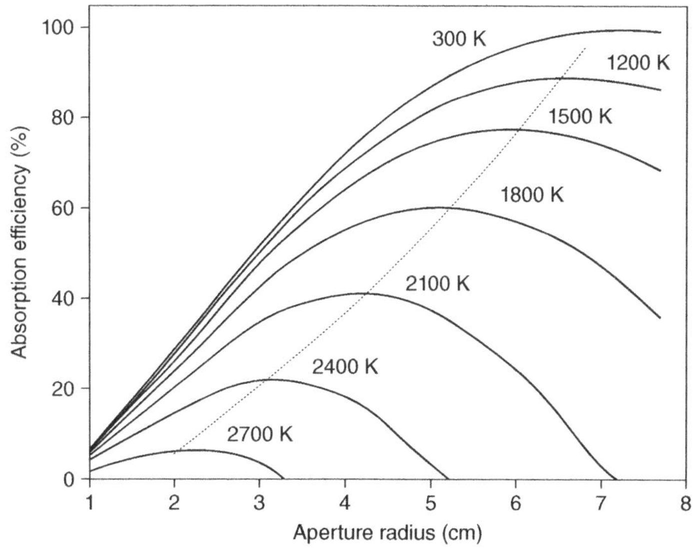

Figure 2.12 Solar flux distribution.

The results are depicted in Figure 2.12 and include a wide range of receiver temperatures and radii. As the aperture size is expanded and more radiation is intercepted, the absorption efficiency initially rises at each operating temperature, reaches a peak, and then falls after the increased radiation loss from the area increase outweighs the additional incident radiation absorbed. There is a best aperture radius for any given operating temperature. Smaller aperture sizes are required by greater temperatures and the concomitant increased quantities of reradiation. The highest absorption efficiency that can be attained is also limited by the use of a higher working temperature.

The results in Figure 2.13 show that an ideal Carnot cycle driven by a black-body receiver on this collector would have a peak overall efficiency when operated at 1,200 K with a 7 cm radius aperture, taking this further to a semi-ideal overall system efficiency by multiplying the absorption efficiency by the Carnot efficiency at the receiver temperature.

2.4.4 Solar Multiple and Capacity Factor

Instead of the design point steady-state efficiency, the total average system efficiency is what matters most. In this regard, the ideas of capacity factor and solar multiple are crucial. A piece of equipment's capacity factor, or CF, is a calculation of its long-term fractional utilization. Full utilization in the case of a CSP system equates to the power that would be generated if the system

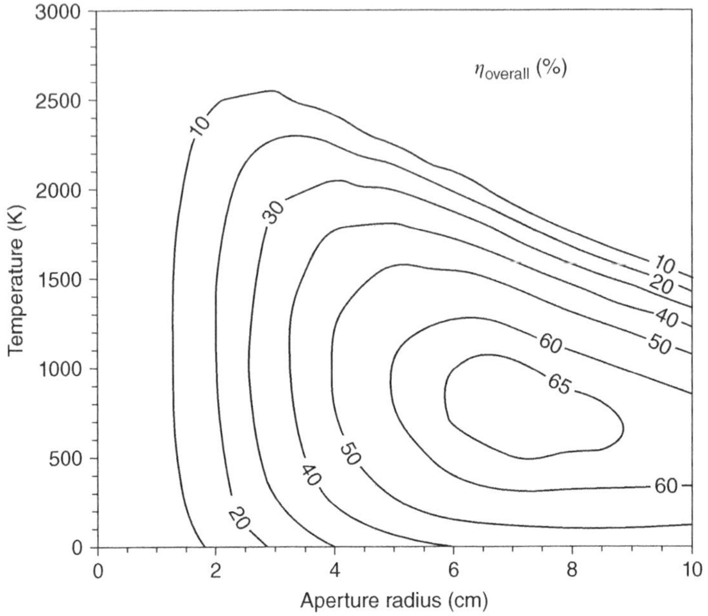

Figure 2.13 Overall system efficiency.

operated at its maximum rated capacity continuously over a given period. The capacity factor (CF) measures the actual output relative to this ideal scenario, reflecting how effectively the system captures and converts solar energy over time, considering variations in solar availability, maintenance, and operational conditions.

Overall system efficiency as a function of the aperture radius and temperature, using the solar flux distribution of Figure 2.13 (reproduced from Steinfeld and Schubnell).

The block operating at the "nameplate" electrical output power $W_{des,pb}$, which is its nominal (rated) design point. If the capacity factor is 100%, the plant is always operating at full capacity; the 25% capacity factor indicates the power block's average electrical output over a long period. 25% of the design point output comes from $W_{avg,pb}$.

$$CF = \frac{W_{avg}}{W_{des.pb}} \tag{2.30}$$

In CSP systems without energy storage or backup boilers, the turbine won't be running at night or when it's cloudy outside; in these systems, the capacity factor will only be 20% as a result of the sun's position and the weather. The capacity factor can be significantly enhanced with the addition of energy storage, up to

75% or more, enabling round-the-clock operation for the majority of the sunnier part of the year.

Oversizing the solar field (adding more heliostats or troughs per steam turbine, for instance) is another way to raise the capacity factor for the power block of a CSP system. This brings us to the definition of the solar multiple, SM, which is the ratio of the thermal power output of the solar field at its design point, Q_{des}, field, which is typically measured at solar noon on a clear midsummer day. The power block's thermal power is needed when operating at full capacity $Q_{des,pb}$.

$$\mathrm{SM} = \frac{\dot{Q}_{\mathrm{des.field}}}{\dot{Q}_{\mathrm{des.pb}}} \tag{2.31}$$

An optimal solar multiple for a system without storage will typically be found to be in the range of 1.15–1.30 when a thorough economic analysis of a CSP plant is conducted. For systems with storage, the optimal solar multiple will typically be found to be 2 or higher, depending on how much storage is installed. The reason for this is that a power block that frequently runs at part-load incurs efficiency and capital cost penalties; it is preferable to waste a small amount of the heat collected during the maximum solar periods in exchange for operating the power block at its design point capacity, at consequently higher efficiency, for a larger portion of the year.

2.4.5 Predicting Overall System Performance

It takes a lot of work to predict a CSP system's output with reasonable precision. Multiple thermally capable subsystem components make up thermal systems, and their behavior at any one time depends not only on the current conditions the entire system is experiencing but also on its previous operational history. Power cycles can take a long time to start up and shut down, and their efficiency fluctuates depending on the load.

Modeling CSP systems can be done in various ways, and research and development in this area is ongoing. One can distinguish between modeling on the half-hour to one-hour time increments and treating the majority of thermal components as being in a "pseudo-steady-state" and more basic approaches that try to follow brief cloud and thermofluid transients.

A recent evaluation of CSP system simulation studies as well as an overview of underpinning theory for the simulation of a whole plant including storage. The Solar PACES organization is actively working to standardize and benchmark CSP simulation tools. Solar thermal applications have made use of IPSEpro, Ebsilon, EcoSimPro, TRNSYS, GateCycle, Dymola, Mathematica, and Aspen, commercial energy system modeling tools. A simple-to-use, open-source, but closed-source program, the System Advisor Model (SAM) from National Renewable Energy

Laboratory (NREL) employs the well-known TRNSYS simulation engine internally. The SOLERGY model of a free and open-source system may be the only one available. Annual power output and capacity factor are important outcomes from such analyses. The study must be combined with an economic analysis to produce optimal sizing and operating strategies.

Results for a trough system developed with the SAM are presented here as an illustration of a system modeling analysis. SAM can forecast the hourly, monthly, and annual production of various alternative renewable energy systems, including CST, CPV, flat PV, and others. There has been a significant corpus of research specifically on its use in CSP systems.

DNI time series data, along with the accompanying ambient temperature, humidity, and wind speeds, are the main inputs for system performance forecasts. SAM accepts weather data in various "typical meteorological year" (TMY) formats, which are artificial years created from actual months from actual years that approximate overall long-term averages and have a representative range of unusual/extreme days. The TMY file is often expected to deliver hourly data.

- For usage with SAM, various templates and case study files with prepared questions are provided. One of them is a confirmed replica of the nearby Las Vegas-area Nevada Solar 1 64 MWe trough installation. The plant's primary characteristics are 64 MWe nominal electrical power output and no storage.
- Total trough aperture (collector) area of 357,428 m^2; solar multiple of 1.264 (i.e. solar field is enlarged about power block system capacity at design conditions).
- The daily profiles of DNI and net generation levels for two days with partial cloud events in the afternoon and one clear day are shown in Figure 2.14. There are several significant outcomes.
- Each day, just before the system starts up, there is a period of negative generation (net energy consumption), as the system is getting ready to run.
- As the system starts up in the morning, generation levels lag behind growing DNI levels, showing system inertia.
- System output can fluctuate during periods of continuous high DNI levels, likely reflecting changes in variables like ambient temperature that affect overall performance.
- When DNI levels fall below a certain point, no generation is created.

The findings for this system's expected monthly output at its Las Vegas, USA, site and three alternate sites in Australia are displayed in Figure 2.15. Table 2.2 includes a list of these scenarios' essential inputs and outputs. The Australian sites in the southern hemisphere peak around the start/end of the calendar year, whereas the northern hemisphere site's output peaks in the middle of the year (Figure 2.16).

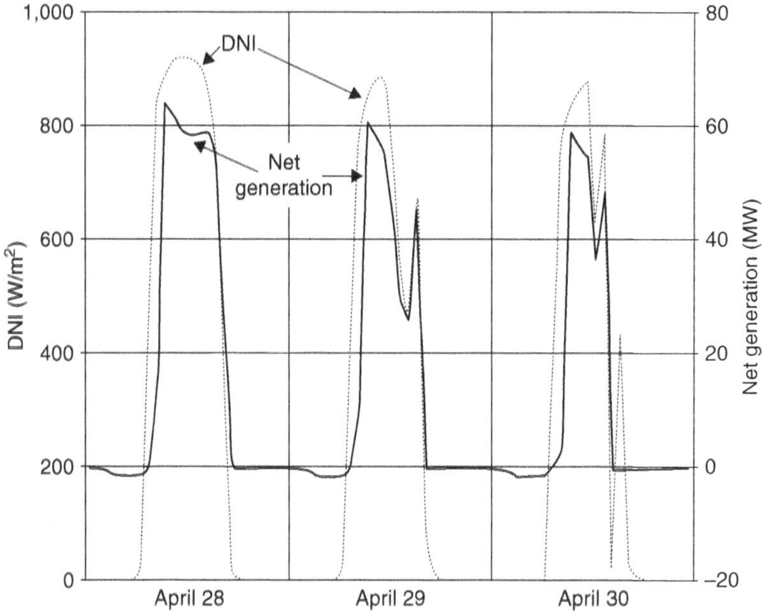

Figure 2.14 Three days of the modeled output of a 64 MWe parabolic trough system using Longreach, Australia, weather data points.

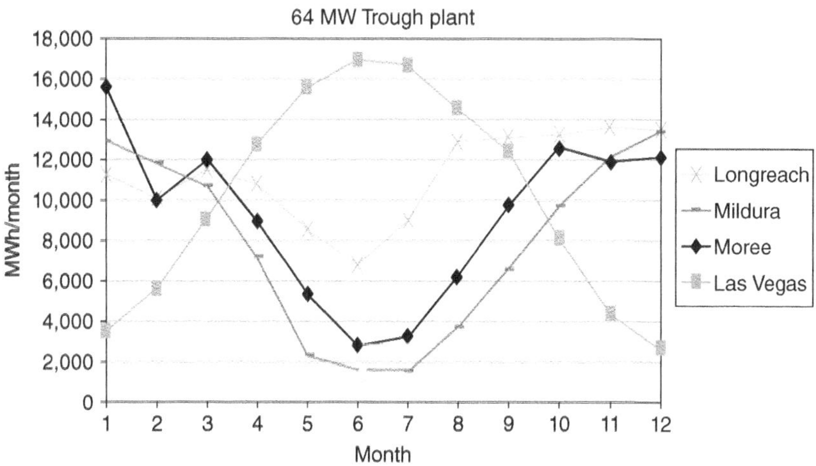

Figure 2.15 The modeled output of a 64 MWe parabolic trough system was sited at Las Vegas (USA) and Longreach, Moree, and Mildura (Australia).

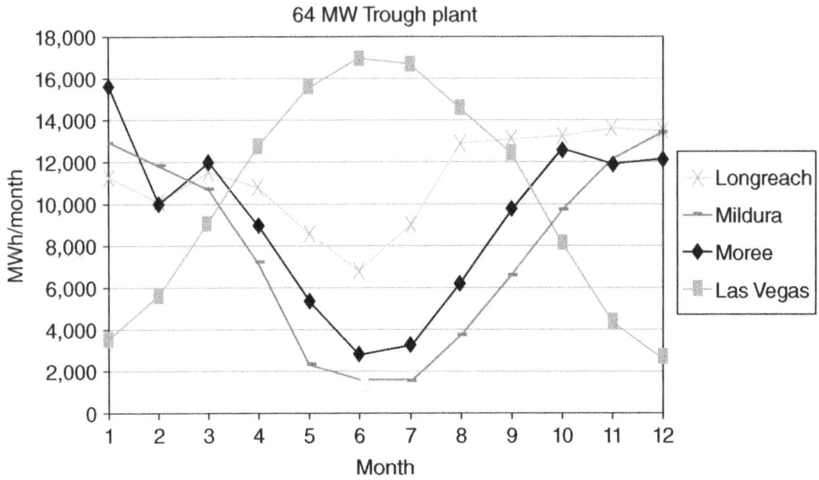

Figure 2.16 Trough plant.

Table 2.2 Average wind speeds.

Location	Net annual generation (MWhe)	Annual conversion efficiency	Capacity factor	DNI (kWh/m^2/yr)	Latitude	Longitude	Average ambient temp. (°C)	Average wind speed (m/s)
Las Vegas (Nevada, USA)	117,147	12.57%	20.90%	2,606.6	36.08	−115.167	13.5	4.1
Longreach (North Queens-land, Australia)	128,794	14.05%	22.90%	2,564.4	−22.43	144.28	22.9	2.4
Moree (NSW, Australia)	106,165	12.17%	13.90%	2,254.6	−23.48	143.83	13.7	2.3
Mildura (North Victoria, Australia)	89,714	3.81%	16.00%	2,124.6	−34.23	142.08	16.9	2.4

The Longreach station, which is closest to the equator, exhibits the most consistent output throughout the year. Annual DNI and total annual generation are correlated, although not linearly. Table 2.2 demonstrates that Longreach outperforms the Las Vegas site despite having a little lower DNI. On the other hand,

Mildura exhibits a 40% reduction in output while having a DNI that is around 20% lower than the top locations. Although performance in Australia's interior is improved by proximity to the equator, it should be noted that equatorial regions often give poor performance due to tropical cloud and humidity.

Average wind speeds might help to lower the system output, but there doesn't seem to be much of a difference in these factors between the sites. The degree to which low DNI days are composed of brief periods of breaking cloud or entire days without sunlight is another differentiator. For CST systems, broken clouds reduce production because it takes too long for the system to achieve its operational temperatures, which renders operation incredibly inefficient.

2.4.6 Economic Analysis

According to the comments above, designing a CSP system should take cost trade-off considerations into account in addition to merely maximizing efficiencies. The necessary elements are introduced in this part to complete the picture.

CSP systems don't require fuel and have high capital expenses. The operating life of a system must be used to amortize initial investments. The net present value (NPV) over the lifetime is therefore the crucial factor. The fundamental equation for calculating NPV is

$$\mathrm{NPV} = \sum_{i=1}^{N} \frac{C_i}{(1 + \mathrm{DR})^i} \tag{2.32}$$

where N is the total number of compounding periods, DR is the discount rate, and C_i are the cash flows that will occur at time interval.

Either nominal or real currency units can be used to measure cash flows. Regardless of the year it occurs, a real cash flow is adjusted for inflation and stated in the currency value of that year and, as a result, has a constant effective worth. There are two types of discount rates: nominal and real. The same NPV can be calculated using nominal currency cash flow data and nominal discount rates or using real currency cash flow measurements and real discount rates.

The main cash flows for a CSP system are the initial capital investments (negative), ongoing operations and maintenance (O&M) costs, costs of ongoing inputs like fuel for hybrid operation or water for cooling (negative), revenue from direct energy sales (positive), and potential provision of ancillary services (positive). The discount rate and the projected plant lifetime are important factors that have a big impact on the final NPV figures. This is because a lower discount rate reduces the extent to which future cash flows are discounted, making them more valuable in present terms. Similarly, a longer plant lifetime extends the period over which the project generates revenue, adding more positive cash flows to the NPV calculation and thereby enhancing the project's overall financial attractiveness. The most common economic performance statistic for power generation plants is the levelized cost of energy (LCOE, also known as levelized energy cost, LEC). It is a common metric applied to all generation technologies, not just CSP and other renewable energy systems. It is described as the constant cost of energy per unit that, throughout the course of the system's lifespan, will produce a total NPV of zero. In other words, it is the energy's continuous sale price that "breaks even."

$$\mathrm{NPV} = \sum_{1}^{N} \left({}^{(\mathrm{LCOE} \times E_{\mathrm{annual}} \times (1-T))} / {}_{(1+\mathrm{DR})^{i}} \right) - \mathrm{NPV}_{\mathrm{LCC}} = 0 \tag{2.33}$$

where $\mathrm{NPV}_{\mathrm{LCC}}$ is the net present value of all life cycle costs, T is the tax rate, and E_{annual} is the annual amount of electrical energy produced.

$$\mathrm{LCOE} = \frac{\mathrm{NPV}_{\mathrm{LCC}}}{\sum_{1}^{N} \left({}^{(E_{\mathrm{annual}} \times (1-T))} / {}_{(1+\mathrm{DR})^{i}} \right)} \tag{2.34}$$

Tax would be excluded from the calculation from a purely social standpoint, but from the perspective of the plant owner or business, it is believed that tax would be imposed on energy sales while other tax deductions would work to reduce costs.

For fossil fuel power plants, the fuel that is continuously consumed in inverse proportion to the conversion efficiency will make up a significant component of the life cycle expenses. The initial capital expenditure makes up the majority of the lifespan costs for CSP and other renewable energy systems, with continuous O&M costs making up a relatively small yet important portion.

Since LCOEs are always reported in dollar values from the year zero, whether they are expressed in real (inflation independent) or nominal terms, this might be confusing. A real LCOE has a constant "value," whereas a nominal LCOE represents a fictitious income that loses real worth with time. The nominal LCOE will be larger because both methods must yield the same total NPV by definition. While nominal LCOEs are frequently utilized for short-term actual projects, real LCOEs are typically employed for long-term technological estimates.

It can be claimed that tax considerations can be excluded from the LCOE from a purely societal standpoint. The general consensus is that, in order for a business organization to break even on a system, it must be assumed that the energy produced is taxed at the usual corporate tax rate. In contrast, tax deductions are available for interest, depreciation, and operational expenses. Complex calculations that add up each discounted cash flow across the system lifespan are the foundation of in-depth, project-specific LCOE evaluations. These calculations are then iteratively solved to determine the real dollar worth of energy, which results in a total NPV of zero. Common problems include the following: debt financing (loans) may be repaid over a longer period of time than equity investments; tax advantages may be available in some jurisdictions but not others; tax-deductible depreciation may be applicable over a shorter period of time than the project; construction is staged over several years and is subject to higher interest rates for financing

- After initial start-up, system output may take some time to settle as commissioning procedures continue.

In addition to general ongoing O&M costs, there may be additional predictable fluctuations in system output over time (such as a component with a known rate of degradation), substantial plant upgrade costs may be anticipated at specific intervals, and various inputs may be subject to varying escalation rates. These problems are all project-specific, dependent on the type of technology, the stage of development, and the site.

Studies that report the LCOE for CSP systems and other generating types frequently do a poor job of thoroughly documenting all of their input parameter assumptions. The methodology is frequently purposefully hidden because it is included in private financial models.

A methodology based on a life cycle costs NPV calculation, represented in the following formula, is considerably simplified but has enough complexity to allow tax, cost of equity, and cost of debt issues to be considered.

$$\begin{aligned} \mathrm{NPV_{LCC}} = \mathrm{EQ} - \sum_{1}^{\mathrm{ND}} \frac{\mathrm{DEP} \times T}{(1+\mathrm{DR})^j} + \sum_{1}^{\mathrm{NL}} \frac{\mathrm{LP}}{(1+\mathrm{DR})^j} - \sum_{1}^{\mathrm{NL}} \frac{\mathrm{INT} \times T}{(1+\mathrm{DR})^j} \\ + \sum_{1}^{N} \frac{\mathrm{AO} \times (1-T)}{(1+\mathrm{DR})^j} - \frac{\mathrm{SV}}{(1+\mathrm{DR})^N} \end{aligned} \tag{2.35}$$

where EQ represents the project developer's initial equity commitment, ND is for the number of years that the system can be depreciated for tax reasons, and DR stands for the nominal discount rate. DEP is the annual depreciation amount, T is the tax rate that applies to the fictitious income from energy sales, LP is the annual loan payment, INT is the annual interest payment that decreases when the loan is repaid, and NL denotes the loan's length (number of years), and AO is the annual operating expense, the end-of-project salvage value, can be computed using fixed and variable maintenance contributions as well as any necessary fuel expenses. N is the project lifetime.

The following are the simplifying presumptions:

- The analysis starts when the plant is first put into service.
- Over the course of the project, a constant annual energy production is expected.
- Since the equity contribution is calculated at the start of year 1, it is assumed that all construction finance costs have been included in it.
- The depreciation of nominal dollars is linear.

- Loan payments are fixed for the whole length of the loan and are made in nominal dollars depending on the amortization of a debt.
- Over the course of a project, nominal annual O&M costs remain unchanged. (This is possibly the most significant since it does not reflect the lumpy investment predicted on component overhaul.)

If tax is not taken into account, the cost of capital (including debt and equity investments) can be wrapped into a single discount rate, and the debt and equity investments can be rolled into a single capital cost. This will further simplify the LCOE and make it easier to understand. When fixed and variable operating expenses are divided, the outcome is

$$\text{LCOE} = \frac{(F_{\text{R}} + \text{O\&M}_{\text{fixed}})C_{\text{o}}}{PF_{\text{c}}} + \frac{C_{\text{fuel}}}{\eta_{\text{conversion}}} + \text{O\&M}_{\text{variable}} \tag{2.36}$$

where P is the system's nominal design point electrical power capacity; F_{c} is its capacity factor (the annual average fraction of nominal capacity that can be achieved); O&M are its operation and maintenance costs, which are divided into those that are proportionate to generation (variable, expressed in the same units as the LCOE) and those that are fixed annual costs (expressed as a percentage of capital cost per year); C_{o} represents the entire cost of capital; C_{fuel} is the cost of fuel per unit of energy utilized in a hybrid system; $\eta_{conversion}$ is the effectiveness of fuel conversion to electricity and is the factor for capital recovery.

The capital recovery factor, which is often referred to as the annualization factor, is dimensionally identical to the discount rate and denotes a rate of repayment that accounts for both the return on investment (at the specified discount rate) and the capital repayment during the system's lifetime. Table 2.3 shows how the capital recovery factor depends on the discount rate and the system life.

All CSP research and development activities are primarily focused on lowering building costs, boosting productivity, and raising capacity factors because these are the available technical means of lowering the LCOE.

2.4.7 Stochastic Modeling of CSP Systems

Investors will demand analysis that is more accurate and detailed as a CSP project gets closer to being financed and built. In other words, it is necessary to be aware of the LCOE's confidence interval. A thorough investigation of the propagation of errors from all measurement and data sources, including estimated values, is necessary to accomplish this. Weather data errors, fluid qualities such as the specific heat capacity of storage medium and HTF, power cycle performance, and collector and pipe-work heat transfer coefficients are a few of these. This type of analysis may now be performed using the SAM simulation tool, and several other simulation programs now provide a similar capability.

Table 2.3 The dependence of capital recovery factor on the discount rate for system lifetimes of 20, 25, and 30 years.

Discount rate (%/yr)	Capital recovery factor for 20-year life (%/yr)	Capital recovery factor for 25-year life (%/yr)	Capital recovery factor for 30-year life (%/yr)
2.00	6.72	5.74	5.10
4.00	7.36	6.40	5.78
5.00	3.02	7.10	6.51
6.00	3.72	7.82	7.26
7.00	3.44	3.58	3.06
3.00	3.19	3.37	3.88
3.00	3.95	3.18	3.73
3.00	3.75	3.02	3.61
3.00	12.56	3.87	3.50
12.00	12.39	12.75	12.41

2.5 The Structure of the Sun

The sun, which has a diameter of 1.399×10^9m and is typically 1.5×10^{11}m away from the earth, is a sphere of extremely hot gaseous matter. The effective black-body temperature of the sun is 5,777 K. Various estimates place the temperature in the core interior region between 8×10^6 and 40×10^6 K, and the density there is thought to be around 100 times that of water. The sun makes up 93.86% of the solar system's mass and is 333,400 times as massive as the earth. It is made up of 2% other elements, 20% helium, and 78% hydrogen. According to estimates, 90% of the energy is produced between 0 and 0.23R (where R is the sun's radius), an area with a density of around 105 kg/m^3 and 40% of the sun's mass. The convective zone extends from 0.7 to 1.0R, where temperatures drop to around 6,000 K and densities range from 10 to 5 kg/m^2. At a distance of 0.7R from the center, temperatures drop to nearly 130,000 K and 70 kg/m^3, respectively.

2.5.1 The Solar Irradiance Spectrum

The wavelength region of 0.25 to 2.0 μm has the highest concentration in the sun irradiance spectrum. Based on wavelength, the supplied solar irradiance spectrum is separated into many bands including the UV, Visible, and IR bands. Only 6% of the total energy is found in the region of 0.2–0.38 m, often known as the UV zone, whereas 48% of it is found in the visible band between 0.38 and 0.78 μm and 46%

is found in the IR band between 0.78 and 2.0 μm. 1367 W/m^2 is the extraterrestrial solar irradiance (I_{sc}). The extraterrestrial radiation flow, on the other hand, varies by about 3% for changes in the earth–sun distance and can be computed using the equation provided below.

$$I_{on} = I_{sc}\left(1 + 0.033\cos\left(\frac{360n}{365}\right)\right) \tag{2.37}$$

where I_{on} is the alien radiation detected on the nth day of the year, measured on the plane normal to the radiation, and I_{sc} is the solar constant.

Solar Constant and Solar Irradiance: At the mean earth–sun distance, the solar constant is the amount of solar irradiance that passes through a unit area at the top of the atmosphere perpendicular to the direction of the irradiance. It has a 1,367 W/m^2 value. The amount of energy per square meter received from the sun in the form of electromagnetic radiation is known as solar irradiance.

Depletion of Solar Radiation by the Atmosphere: The atmosphere that surrounds the globe is made up of different gases, suspended dust and other particles, water vapor, and different kinds of clouds. The solar radiation is partially absorbed, dispersed, and reflected in distinct wavelength bands as it travels through the atmosphere. Certain wavelengths of irradiance are absorbed by water vapor, ozone, CO_2, and oxygen. In a process called Rayleigh scattering, where the intensity is inversely proportional to the fourth power of the light's wavelength, radiation is dispersed by gas molecules and tiny dust particles. Mie Scattering will occur if the particle size is greater than the wavelength of light. Because of clouds, bigger particles, and other materials in the atmosphere, radiation will be reflected. Numerous different forms of clouds also absorb a substantial quantity of solar radiation. Due to dispersion, reflection from the ground, and clouds, some irradiance is reflected into the atmosphere. The albedo of the ground is the percentage of incident light that is reflected, and it is typically 0.2. Direct solar irradiance, also known as beam irradiance, refers to the solar radiation that reaches the surface of the globe. Diffused solar irradiance is the term used to describe the solar radiation that returns to the earth's surface in the short wavelength region after being reflected, absorbed, or scattered but is not entirely lost in the atmosphere. Global or total solar irradiance refers to the sum of diffuse and direct irradiance on the earth's surface.

2.5.2 Factors Affecting the Availability of Solar Energy on a Collector Surface

- Geographical area
- Collector's position on the site
- Orientation and tilt of the collector
- Daytime, season, and atmospheric conditions
- Kinds of collectors

2.6 Radiation Instruments

- Pyranometer
- Pyrheliometer
- Pyrgeometer
- Net radiometer
- Sunshine recorder

The sensitivity, stability, and accuracy of these devices determine whether they are first class, second class, or third class.

2.6.1 Solar Irradiance Components

Solar radiation that reaches the surface of the planet directly is known as direct radiation. Molecular and aerosol particles scatter as they enter the earth's atmosphere. Global solar radiation is equal to the sum of direct and diffuse solar radiation. In addition to using direct radiation, concentrators also use some dispersed radiation. Flat-plate collectors make use of reflected solar radiation in addition to direct and diffuse solar radiation.

2.6.2 Instruments Used

On a horizontal surface, the global sun irradiance is measured using a pyranometer. The measurement of diffuse sun irradiance is done with a pyranometer with a shade ring. A pyrheliometer is used to measure the direct radiation from the sun.

2.6.3 Detectors for Measuring Radiation

Calorimetric sensors track the radiant energy that strikes a high-conductivity metal that has been painted with a nonselective, very absorbent black paint. Thermomechanical sensors bend a bimetallic strip to monitor radiant flux. These sensors are made up of two different metallic wires joined at the ends. The field of measuring sun radiation uses the most photoelectric sensors, such as PV devices. A semiconducting substance, such as silicon, is used to create PV devices.

2.6.4 Measuring Diffuse Radiation

Diffuse radiation is measured using the same equipment as is used to calculate total or global radiation. The receiver (pyranometer) is shielded from direct sun radiation using an appropriate device (a disk or shadow ring).

The accuracy of the tool is impacted by the following elements:

The precision of the measurement is impacted by many reflections inside the glass cover. It is assumed that the sky is isotropic when calculating the correction factor. The shade device also stops a portion of solar energy from reaching the receiver. The receivers' dimensions are not sufficiently standardized.

2.7 Why Solar Energy Estimation?

The information on the hourly worldwide sun irradiance is crucial for numerous uses, including the production of solar power. Such information is similarly crucial for designing energy-efficient structures. The community at large as well as the locality where the irradiance statistics are obtained can benefit from information on the global sun irradiance received at any station. Solar energy has a vast potential and might be used for various purposes, including the production of electricity. In a nation like India, where millions of people live in rural places and are unable to access the grid supply, such potential offers the chance to create power on a dispersed basis. Even for those locations where measurements have already been made, hourly data on global sun radiation measurements are not available in many nations, including India. Although exact measurements are necessary to meet the demand for energy from this environmentally favorable and pure source of energy, they are rarely available in many poor nations due to the cost and labor involved. As a result, it is crucial to suggest a powerful substitute for a solar radiation estimator that is based on more easily accessible meteorological factors. Since there are so few stations in India that measure solar irradiance, it is usual practice to estimate solar irradiance from easily quantifiable metrological characteristics like sunshine length, temperature, etc. A superior method for determining the sun irradiance of areas where measurements have not been made is accurate modeling, which depends on the caliber and volume of the measured data used. In the literature, there are several methods for calculating solar energy, including mathematical, regression, and intelligent models based on wavelet transformation, neural networks, fuzzy logic, etc. This chapter discusses a handful of these models.

2.8 Mathematical Models of Solar Irradiance

In many solar energy applications, such as PV and solar building heating, a correct calculation of solar energy is crucial. To accurately estimate solar energy, beam, and diffuse energy on a horizontal surface should be determined. On the other hand, the amount of solar energy at the top of the atmosphere is sufficiently known. One can forecast terrestrial beam and diffuse energy if one can properly understand how solar energy is scattered and absorbed by various elements of the environment. These are the various solar energy attenuation phenomena that are continuously improving in understanding and leading to new estimating methods. The capacity to analyze air components like ozone, aerosols, and water vapor, which scatter and absorb solar radiation, has also seen constant progress. This all improves the estimation's accuracy. Even if measured data have spatial and temporal restrictions, they are nevertheless the best data available. Even in developed nations, there are many locations where data are not measured; as a result, the typical metrological year (TMY) and solar energy atlas for these stations are based on modeled data.

2.8.1 CPCR2 (Code for Physical Computation of Radiation, 2 Bands) Model

This model calls for commonly accessible input parameters and contains the parameterization of extinction mechanisms. The sun spectrum is split into two bands in this two-band, clear-sky solar energy modeling technique: an ultraviolet/visible band, or B1 (0.29–0.7 μm), and an infrared band, or B2, which ranges from 0.7 to 2.7 μm. The World Radiometric Center (WRC)'s alien solar spectrum is used in the model, with a solar constant value of 1,367 Wm^{-2}. The CPCR2 model accounts for 46.04% and 50.57%, respectively, of the solar constant in the UV/visible and infrared bands. As a result, bands B1 and B2 of the total extraterrestrial energy, E_{on1}, and E_{on2}, respectively, should be given the coefficients $f_1 = 0.4604$ and $f_2 = 0.5057$.

Direct Normal Energy: The below equation can be used to calculate the beam solar energy:

$$E_{\text{bn}i} = T_{\text{R}i} T_{\text{g}i} T_{\text{o}i} T_{\text{w}i} T_{\text{a}i} E_{\text{on}i} \tag{2.38}$$

where E_{bn} is the beam solar energy at the normal surface (W/m^2), T_{R} is the Rayleigh transmittance (dimensionless), T_{g} is the transmittance due to mixed gases (dimensionless), T_{o} is the ozone transmittance, T_{w} is the transmittance of water vapor, T_{a} is the transmittance of aerosols (dimensionless), and $i = 1$ for band B_1 and $i = 2$ for band B_2.

The equation calculates the total energy of the beam at the ground level.

$$E_{\text{bn}} = E_{\text{bn1}} + E_{\text{bn2}} \tag{2.39}$$

2.9 Diffuse and Global Energy

The three separate components that represent the two scattering layers (molecules and aerosols) and the process of backscattering between the ground and sky are combined to represent the diffuse energy at the ground level. The equation can be used to calculate the diffuse energy.

$$E_{\text{d}i} = E_{\text{dR}i} + E_{\text{dA}i} + E_{\text{dd}i} \tag{2.40}$$

where E_{d} represents diffuse energy (W/m^2), E_{dR} represents molecular scattering (W/m^2), E_{dA} represents aerosol scattering (W/m^2), and E_{dd} represents backscattering between the ground and sky (W/m^2).

The equation gives the total energy:

$$E_g = E_d + E_b \tag{2.41}$$

2.10 REST2 (Reference Evaluation of Solar Transmittance, 2 Bands) Model

When compared to more straightforward broadband models, the CPCR2 two-band model performed best, according to earlier findings from in-depth performance assessment tests. A recent comprehensive investigation showed that CPCR2 consistently performed effectively in predicting direct normal energy (DNI) under both ideal and practical circumstances. The current contribution outlines several novel aspects of the model, highlights recent algorithmic advancements, and suggests a benchmark dataset for the evaluation of this or any other model of a similar nature. With a band spacing of 0.7 μm, REST2's overall structure is nearly identical to that of CPCR2. Band 1 encompasses visible and UV light between 0.29 and 0.70 μm. It is distinguished by significant ozone UV absorption and strong molecule and aerosol scattering over the whole band. Water vapor, carbon dioxide, and other gases absorb strongly in Band 2, which spans the near-infrared spectrum from 0.7 to 4 μm, with very little scattering. It has been demonstrated that this method, which has been applied to a few other models in the literature, has two intriguing benefits: (1) it increases accuracy in comparison to standard single-band models and (2) it makes the process of calculating energy, whose spectral ranges almost exactly match band 1, simpler. The extra-atmospheric energy at the mean sun–earth distance, calculated using the most recent extraterrestrial spectral energy distribution and the most recent solar constant value of 1366.1 W/m^2, is E_{on1} = 635.4 W/m^2 (or 46.51%) and E_{on2} = 703.7 W/m^2 (or 51.95%) in the two bands, respectively. The parameterizations for direct and diffuse energy, in contrast to CPCR2, allow for the circumsolar energy occupied by the typical entire field of view of tracking devices.

2.11 Direct Energy

The formalism is substantially the same as in CPCR2, except for the addition of a nitrogen dioxide absorption provision, similar to that in REST. The band direct normal energy, E_{bni}, for each of the two bands, i, is created by multiplying the separate transmittances:

$$E_{bni} = T_{Ri}T_{gi}T_{oi}T_{ni}T_{wi}T_{ai}E_{oni} \tag{2.42}$$

where T_{Ri}, T_{gi}, T_{oi}, T_{ni}, T_{wi}, and T_{ai} are the band transmittances for, in that order, Rayleigh scattering, uniformly mixed gas absorption, ozone absorption, nitrogen dioxide absorption, water vapor absorption, and aerosol extinction.

By adding the two-band components, the broadband DNI may be calculated easily:

$$E_{bn} = E_{bn1} + E_{bn2} \tag{2.43}$$

2.12 Diffuse and Global Energy

The formalism used here is similar to CPCR2 in that it is based on a two-layer scattering technique. It is hypothesized that all ozone and mixed gas absorption, as well as all Rayleigh scattering, originate from the top layer. Similar to this, it is believed that all aerosol scattering, as well as aerosol, water vapor, and nitrogen dioxide absorption, originates from the bottom layer. The down-welling diffuse energy is considered to act as direct energy at an effective air mass, $m = 1.66$ after scattering has taken place in the top layer. This is the amount of air mass utilized to determine the transmittances relating to bottom-layer absorption.

If a ground has zero albedo and is perfectly absorbing, the incident diffuse energy is defined as follows:

$$E_{dpi} = T_{oi}T_{gi}T_{oni}T_{owi}\left[B_{Ri}(1 - T_{Ri})T_{ai}^{0.25} + B_a F_i T_{Ri}\left(1 - T_{ai}^{0.25}\right)\right]E_o \tag{2.44}$$

where

$$E_{oi} = E_{oni}\cos Z \tag{2.45}$$

where the zenith angle Z is used. Function F_i is a correction factor that was added to the straightforward transmittance method employed here to account for multiple scattering effects and other drawbacks. The forward-scattering percentages for the Rayleigh extinction are B_{R1} and B_{R2}. They would be exactly 0.5 in the absence of multiple scattering since molecules scatter equally in both the forward and backward directions. Band 2 has little multiple scattering ($B_{R2} = 0.5$), however, Band 1 does not. After spectral integration and parameterization, B_{R1} is calculated as follows, using a straightforward spectral model to represent this effect:

$$B_{R1} = 0.5\left(0.89013 - 0.0049558m_R + 0.000045721m_R^2\right) \tag{2.46}$$

As in CPCR2, the aerosol forward scattering factor, B_a, is the same

$$B_a = 1 - \exp(-0.6931 - 1.8326\cos Z) \tag{2.47}$$

The resulting figures for the broadband energy incident on an ideal-black, horizontal surface are as follows:

$$\begin{aligned} E_b &= E_{bn}\cos Z, \\ E_{dp} &= E_{dp1} + E_{dp2}, \text{ and} \\ E_{gp} &= E_b + E_{dp} \end{aligned} \tag{2.48}$$

Under typical circumstances, a backscattered contribution must be added due to the interaction between the earth's surface, which reflects light, and the atmosphere's scattering layers. This contribution is often negligible about 10% of E_{gp} but it could become significantly more significant in snowy areas. The ground albedo that needs to be taken into account in this case, gi, is an average across a 5–50 km circle surrounding the place in question. The backscattered diffuse component, E_{ddi}, is created by taking into account various reflections between the ground and the atmosphere, and the sky albedo, ρ_{si}, is acquired for each band as a function of α_i and β_i (angstrom turbidity coefficient [dimensionless]):

$$E_{ddi} = \rho_{gi}\rho_{si}(E_{bi} + E_{dpi})/(1 - \rho_{gi}\rho_{si}) \tag{2.49}$$

where $E_{bi} = E_{bni}\cos Z$. The broadband diffuse energy is acquired as $E_d = E_{d1} + E_{d2}$ and the broadband global energy is derived as $E_g = E_b + E_d$. Finally, the total diffuse energy in each band is calculated as $E_{di} = E_{dpi} + E_{ddi}$.

2.12.1 Reference Evaluation of Solar Transmittance Model

This model was suggested by Gueymard. The total absorption of NO_2 is not considered in this model. The model's suggested shape is as follows:

$$E_{bn} = E_{on}T_R T_g T_o T_w T_a \tag{2.50}$$

E_{bn} stands for direct normal irradiance, E_{on} for extraterrestrial irradiance, T_R for Rayleigh transmittance, T_o for ozone, T_g for uniformly mixed gas, T_w for water vapor, and T_a for aerosol transmittance.

2.12.2 Estimation of Global Irradiance

The following can be expected for the worldwide solar irradiance:

$$E_g = E_{bn}\cos z + I_d \tag{2.51}$$

where E_g is the expected global sun irradiance at the horizontal surface, E_d is the expected diffuse solar irradiance at the horizontal surface, and E_{bn} is the computed beam irradiance at normal incidence.

2.12.3 Estimation of Diffuse Irradiance

On the basis of the parametric model proposed by Iqbal, the diffuse irradiance is computed. The horizontal diffuse irradiance E_d at the ground level is a combination of three distinct components, which correspond to the Rayleigh scattering after the first pass through the atmosphere, D_r; the aerosols scattering after the first pass through the atmosphere, D_a; and various reflection processes between the ground and sky, D_m:

$$E_d = D_r + D_a + D_m \tag{2.52}$$

2.13 Regression Models

There are other models based on linear, multiple quadratic, etc., regression methodologies. In this part, a few regression models for the Indian scenario are described. Any scenario's model can be created using the same method. For seven Indian stations—New Delhi, Nagpur, Ahmedabad, Jodhpur, Kolkata, Vishakhapatnam, and long—the monthly average solar irradiance, sunshine duration hours, temperature, and other data on meteorological parameters are obtained from various sources, including the National Institute of Solar Energy and the Indian Meteorological Department (IMD), New Delhi. Based on their geographic and climatic zones, these were taken into consideration. Table 2.4 displays the geographic characteristics of the Indian stations taken into consideration for this analysis.

Tables 2.4 and 2.5 exhibit the monthly average values for S/S_o and H/H_o for Jodhpur and New Delhi, respectively. Tables 2.6 and 2.7, respectively, present the estimates of diffuse irradiance values of S/S_o and H_d/H_g for New Delhi and Jodhpur.

The trends of the global and diffuse solar irradiance in Jodhpur and New Delhi, respectively, for various months of the year, are shown in Figures 2.17 and 2.18 (Table 2.8).

Many writers have tried to establish correlations between the values of solar irradiance (global or diffuse) and climatic variables such as the number of sunny hours, cloud cover, and precipitation because measurements of solar irradiance are frequently not accessible. To differing degrees, the different meteorological factors are related to sun irradiation. One can determine a correlation between these two variables at the few meteorological stations that record both irradiance and sunshine duration.

The original Angstrom model, which served as the basis for this modeling. It takes the following form:

$$\frac{\overline{H}}{H_o} = a + b\frac{\overline{S}}{S_o} \tag{2.53}$$

Table 2.4 Geographical features of the stations in India considered in this work.

Station latitude (0N)	Longitude (0E)	Height above mean	Sea level (m)	Climate zone
New Delhi	23.58	77.20	216	Composite
Nagpur	21.10	73.05	31	Composite
Ahmedabad	22.07	72.63	55	Hot and dry
Jodhpur	26.30	72.02	224	Hot and dry
Kolkata	22.65	83.45	6	Warm and humid
Vishakhapatnam	17.72	82.23	3	Warm and humid
Shillong	25.57	91.88	1600	Cold and cloudy

Table 2.5 Monthly average data of S/S_0 and H/H_0 for Jodhpur, India.

Month	S/S_0	H (MJ/m^2)	H_0 (MJ/m^2)	H/H_0
January	0.89	15.89	22.98	0.787
February	0.89	13.09	23.18	0.642
March	0.80	21.16	32.96	0.642
April	0.89	22.57	37.07	0.608
May	0.89	22.56	32.29	0.599
June	0.83	22.73	33.94	0.570
July	0.72	13.21	33.49	0.486
August	0.73	17.91	37.78	0.474
September	0.81	13.84	34.34	0.577
October	0.87	13.35	23.50	0.656
November	0.90	16.68	24.92	0.670
December	0.90	15.04	22.72	0.662

where S is the monthly mean of the daily hours of bright sunshine, S_0 is the maximum daily hours of sunshine (or day length), H is the monthly mean of the daily global irradiance on the horizontal surface, H_0 is the extraterrestrial solar irradiance on the fifteenth of the month, and a and b are the regression coefficients. Equation (2.53) and, respectively, are the suggested models for

Table 2.6 Monthly average data of S/S_0 and H/H_0 for New Delhi, India.

Month	S/S_0	H (MJ/m^2)	H_0 (MJ/m^2)	H/H_0
January	0.836	12.32	20.18	0.660
February	0.857	16.42	22.76	0.691
March	0.857	20.64	31.18	0.662
April	0.868	24.07	40.52	0.594
May	0.878	24.43	45.66	0.535
June	0.847	22.54	45.81	0.492
July	0.697	13.07	33.64	0.481
August	0.697	17.79	33.01	0.456
September	0.847	13.90	37.64	0.502
October	0.900	16.80	22.17	0.725
November	0.853	14.13	17.86	0.791
December	0.804	3.93	16.63	0.717

Table 2.7 Monthly average data of S/S_0 and H_d/H_g for Jodhpur, India.

Month	S/S_0	H_d (MJ/m^2)	H_g (MJ/m^2)	H_d/H_g
January	0.89	4.23	15.89	0.266
February	0.89	5.33	13.09	0.295
March	0.80	6.67	21.16	0.315
April	0.89	7.97	22.57	0.353
May	0.89	3.56	22.56	0.406
June	0.83	3.77	22.73	0.474
July	0.72	3.90	13.21	0.620
August	0.73	3.00	17.91	0.614
September	0.81	7.43	13.84	0.374
October	0.87	13.35	23.50	0.660
November	0.90	16.68	24.92	0.670
December	0.90	15.04	22.72	0.662

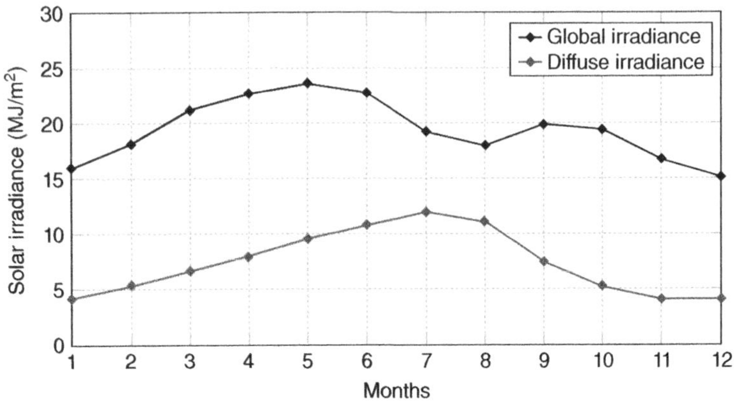

Figure 2.17 Global and diffuse solar irradiance at Jodhpur.

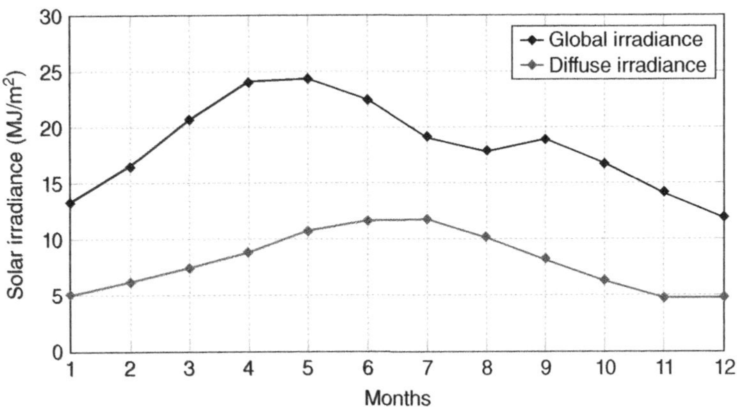

Figure 2.18 Global and diffuse solar irradiance at New Delhi.

estimating the global and diffuse solar irradiance for Jodhpur, India.

$$\frac{\overline{H}}{\overline{H_0}} = 0.3 + 0.4556\frac{\overline{S}}{\overline{S_0}} \tag{2.54}$$

$$\frac{\overline{H_d}}{\overline{H_g}} = 1.45 - 1.62\frac{\overline{S}}{\overline{S_0}} \tag{2.55}$$

The ratio of H/H_0 is the atmospheric transmission coefficient, also referred to as the clearness index, and S/S_0 is the percentage of the maximum number of brilliant sunshine hours. Cooper's formula can be used to calculate S_0's value.

$$\overline{S}_0 = \left(\frac{2}{15}\right) w \tag{2.56}$$

Table 2.8 Monthly average data of S/S_o and H_d/H_g for New Delhi, India.

Month	S/S_o	H_d (MJ/m^2)	H_g (MJ/m^2)	H_d/H_g
January	0.836	5.21	12.32	0.391
February	0.857	6.22	16.42	0.379
March	0.857	7.56	20.64	0.366
April	0.868	3.83	24.07	0.369
May	0.878	3.68	24.43	0.437
June	0.847	3.66	22.54	0.571
July	0.697	3.83	13.07	0.620
August	0.697	3.27	17.79	0.577
September	0.847	3.27	13.90	0.437
October	0.900	6.37	16.80	0.380
November	0.853	4.92	14.13	0.348
December	0.804	4.87	3.93	0.410

The following expression yields the alien solar irradiation on a horizontal surface H_0

$$H_0 = \frac{24 \times 2.6 \times 10^{-3} \times I_{SC}}{\pi}\left(1 + 0.033\cos\left(360\frac{\overline{D}}{365}\right)\right)$$
$$\cos\varphi\cos\delta\sin\omega + w\sin\varphi\sin\delta \tag{2.57}$$

$$\delta = 22.45\sin\left(360\frac{248 + \overline{D}}{365}\right) \tag{2.58}$$

$$\omega = \cos^{-1}(-\tan\varphi\tan\delta) \tag{2.59}$$

where D is the number of the day, $I_{sc} = 1367\,\text{Wm}^2$ is the solar constant, is the location's latitude, is the declination angle, and is the angle of the sun's setting.

The MATLAB® application can be used to calculate an estimate of the alien solar irradiance H_0 and the relative sunshine duration at various Indian stations. The average day of the month is taken into account in computations. The day with extraterrestrial solar irradiance closest to the average for the month is considered the average day.

2.14 Intelligent Modeling

There are several models available for calculating sun irradiance under a clear sky. The regression model and three broadband models—REST, REST2, and

CPCR2—are described in earlier sections. Depending on the location, these models are not suitable for calculating solar irradiance during the monsoon season (June to September) or when there are clouds in the sky. The sections before this one showed a regression model to calculate solar energy using sunshine duration as an input. Although the results were satisfactory, they are only useful under clear-sky conditions. However, a model based on fuzzy logic, artificial neural networks (ANNs), and generalized neural networks is described in the following sections due to the ambiguity of weather conditions. Dust, moisture, aerosols, clouds, and changes in lower atmosphere temperature can all contribute to atmospheric uncertainty. Clouds have the potential to have the most negative effects on the amount of alien solar irradiance that reaches the earth's surface at sea level. On exceptionally clear days, the atmosphere reduces the extraterrestrial solar intake by about 30%; on very cloudy days, it increases by nearly 100%.

It is quite challenging to accurately estimate the outcomes of solar irradiance using a mathematical or regression model in India because there are between 50 and 100 cloudy days every year. Therefore, depending on the availability of this information, fuzzy logic and a neural network-based model may be preferable for determining the solar irradiance at a given location using sunshine duration, temperature, and other meteorological parameters.

2.15 Fuzzy Logic-Based Modeling of Solar Irradiance

Seven Indian stations have been taken into consideration in this part based on their climatic conditions. IMD measures a number of climate variables in India, including solar irradiance, sunshine hours, temperature, rainfall, air pressure, and humidity. However, not all stations are measuring these variables, especially sun irradiance. Since the majority of the stations are very new, they have not gathered extensive data. The IMD, the National Institute of Solar Energy, and the Ministry of New and Renewable Energy (MNRE) provided the raw data for the proposed study via personal conversation. Only two easily measurable quantities have been used as inputs in this section, and it is expected that all other values will remain constant. Each parameter, however, has a unique impact on solar irradiance, and ignoring them results in some estimation error. As a result, all the uncertainties and model issues connected to solar irradiance are taken into account using a straightforward technique in this analysis and fuzzy sets are used to treat the length and warmth of the sunshine linguistically. There are neither assumptions nor model constants in the fuzzy method. In order to reduce mistakes in solar irradiance calculation, this approach was developed. In the current investigation, seven Indian stations are chosen to estimate monthly mean solar irradiance based on sunshine duration and temperature: New Delhi, Kolkata, Mumbai,

Jodhpur, Pune, Ahmedabad, Vishakhapatnam, and Shillong. The solar irradiance is related to sunshine duration and temperature using the MATLAB toolbox (FIS). The inputs are fuzzified into five subsets for each of the stations, including very low, low, medium, high, and very high quantities of input data. But every piece of information used in this research is presented in its normalized form. Similar to this, the five groups of sun irradiation values are categorized according to increasing magnitude. The construction of a fuzzy-based model for solar energy estimation in the MATLAB toolbox has been done using the technique listed below. Figure 2.19 shows the fuzzy inference system editor. The definition of membership functions connected to each of the variables is the following stage. The tool used to display and edit every membership function for the entire FIS, including input and output variables, is called the membership function editor. For the input and output variables in this work, a triangular membership function has been selected. Figure 2.20 through Figure 2.22, respectively, depict the membership functions for temperature, global irradiance, and sunshine duration.

In the proposed study, all of the data are expressed in their normalized form. Table 2.9 lists membership functions and their ranges. A set of multiple-antecedent fuzzy rules has been developed for the estimation of solar irradiance at a specific station. The ratio of temperature and sunshine duration is the rules' inputs, and their resulting output is the solar irradiance shown in Table 2.10. The dark area of the matrix contains the rules' consequences.

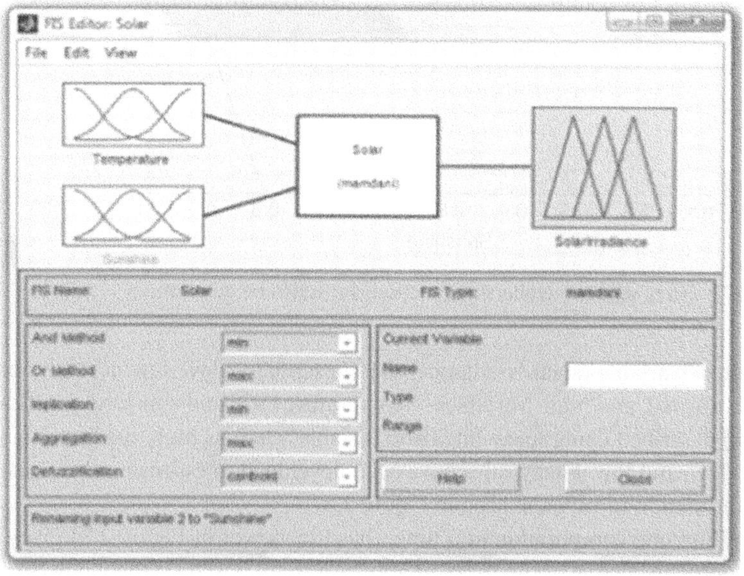

Figure 2.19 Fuzzy inference system editor.

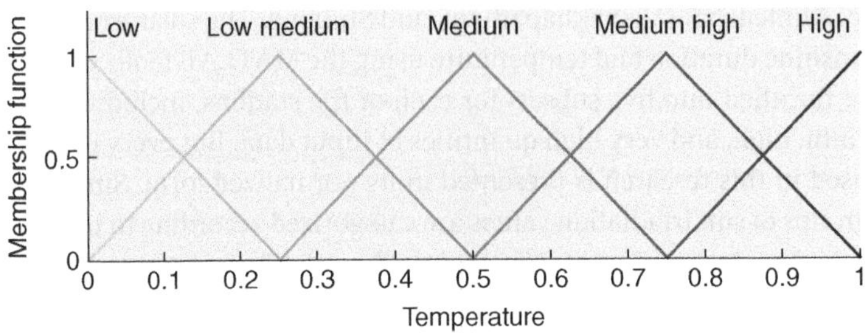

Figure 2.20 Fuzzy subset membership functions for temperature.

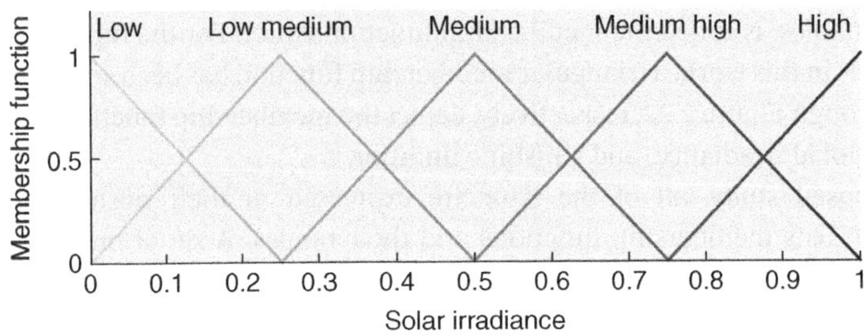

Figure 2.21 Fuzzy subset membership functions for solar irradiance.

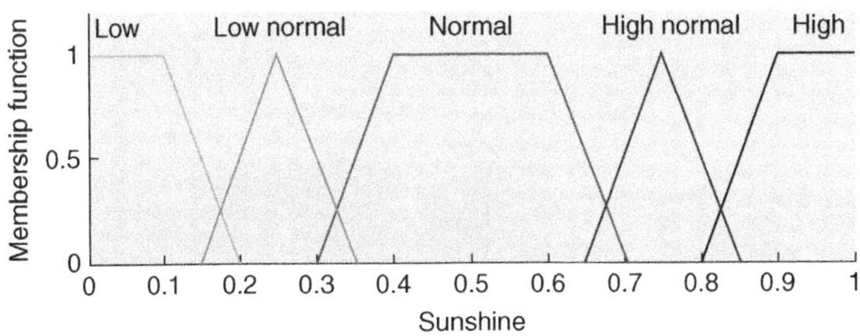

Figure 2.22 Fuzzy subset membership functions for sunshine duration.

The output variable (solar irradiance) is represented as low, low normal, normal, high normal, and high. Variables—temperature ratio and sunshine duration ratio—are described using fuzzy linguistic variables such as high, medium high, medium, low, and low. Fuzzy rules have been developed to estimate solar irradiance at any station in India, as seen in Figure 2.23.

By detailing one computation at a time, the rule viewer in Figure 2.24 gives a type of microperspective of the fuzzy inference system. Based on the whole span of the input set, the surface viewer displays the system's entire output surface.

Table 2.9 Membership functions and range of parameters.

Input				Output	
Membership function	T/T_o	**Membership function**	S/S_o	**Membership function**	H/H_o
Low	[0.00–0.25]	Low	[0.00–0.20]	Low	[0.00–0.20]
Low-Med	[0.00–0.50]	Low-Nor	[0.15–0.35]	Low-Nor	[0.15–0.35]
Medium	[0.25–0.75]	Normal	[0.30–0.70]	Normal	[0.30–0.70]
Med-High	[0.50–1.00]	High-Nor	[0.65–0.85]	High-Nor	[0.65–0.85]
High	[0.75–1.00]	High	[0.80–1.00]	High	[0.80–1.00]

Table 2.10 Decision matrix for determining global solar irradiance.

AND		S/S_o				
		Low	**Low-Nor**	**Nor**	**High-Nor**	**High**
Temperature	Low	Low-Med	Low-Med	Low	Low-Med	Med
	Low-Med	Medium	Low-Med	Low-Med	Low-Med	Med
	Medium	High-Med	Med	Low-Med	Low-Med	High-Med
	Med-High	High-Med	High-Med	Med	High-Med	High-Med
	High	High	High-Med	Med	High-Med	High

Figure 2.25 gives a description of the fuzzy model's three-dimensional surfaces of output.

2.15.1 Datasets

On the basis of sunshine duration and temperature, the monthly mean solar irradiance at the seven Indian sites is calculated. On the basis of absolute relative error (ARE), the estimated values are contrasted with the measured values. For the New Delhi and Jodhpur stations, Table 2.11 provides the monthly average values of the clearness index, which is the ratio of the average daily solar irradiance to the daily maximum irradiance, sunshine ratio (i.e. the ratio of the average daily sunshine hours and the theoretical sunshine duration), and temperature ratio.

The ARE obtained for the New Delhi and Jodhpur stations in the estimation of the diffuse and global solar irradiance is shown in Tables 2.12 and 2.13, respectively. These tables make it evident that ARE varies from 2.6% to 7.4%

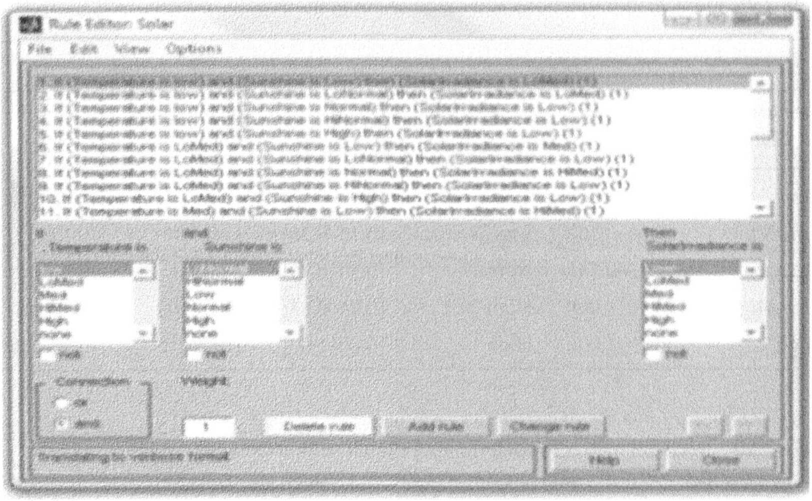

Figure 2.23 Fuzzy rules.

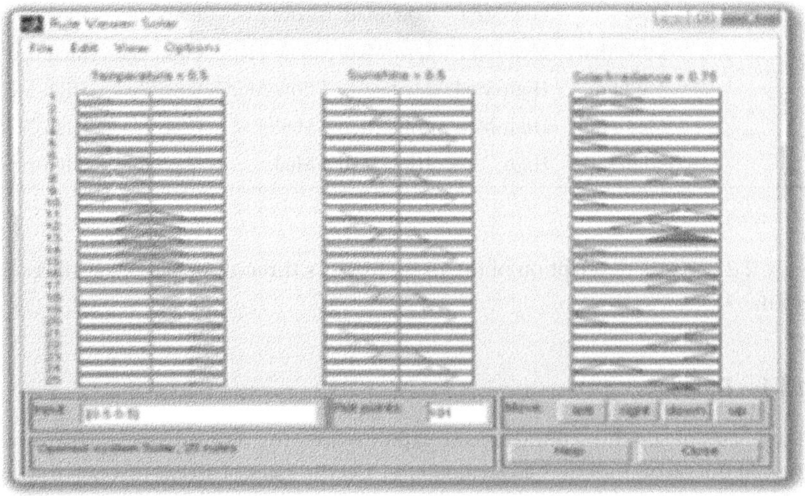

Figure 2.24 Fuzzy rules viewer.

when used to estimate global solar irradiance, but only from 6.3% to 3.35% when used to estimate diffuse solar irradiance. Because fuzzy logic is an approximation method, the obtained ARE in the estimation of solar irradiation is discovered within the required bounds. Results from the fuzzy logic model are also graphically displayed. The observed and predicted global sun irradiance levels for New

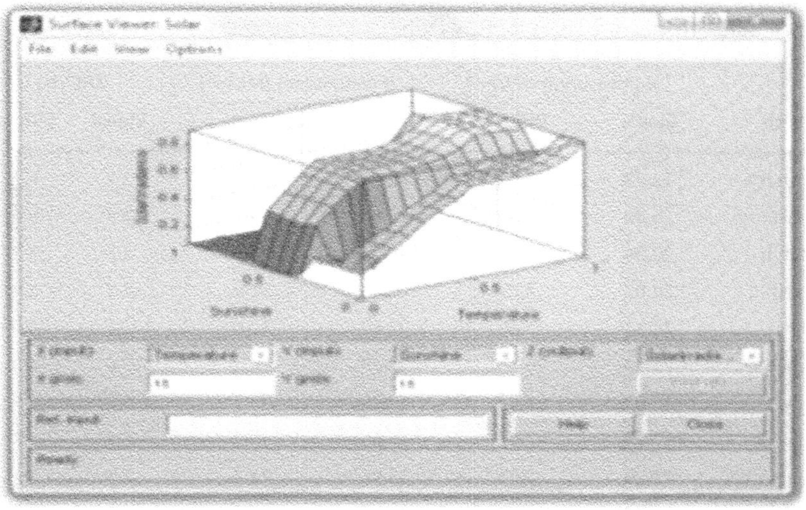

Figure 2.25 Three-dimensional surfaces of output of the fuzzy model.

Table 2.11 Monthly average values of H/H_0, S/S_0, and T/T_0 for New Delhi and Jodhpur.

	New Delhi			Jodhpur		
Month	H/H_o	S/S_o	T/T_o	H/H_o	S/S_o	T/T_o
January	0.189	0.836	0.334	0.180	0.89	0.50
February	0.387	0.857	0.509	0.386	0.89	0.60
March	0.657	0.857	0.639	0.675	0.80	0.76
April	0.877	0.868	0.809	0.807	0.89	0.91
May	0.900	0.878	0.896	0.900	0.89	1.00
June	0.779	0.847	0.900	0.822	0.83	0.99
July	0.557	0.697	0.841	0.492	0.72	0.90
August	0.475	0.697	0.813	0.370	0.73	0.86
September	0.546	0.847	0.673	0.551	0.81	0.84
October	0.412	0.900	0.727	0.505	0.87	0.81
November	0.241	0.853	0.596	0.254	0.90	0.65
December	0.100	0.804	0.464	0.100	0.90	0.54

Table 2.12 ARE percentages in the estimation of monthly mean solar irradiance in comparison with measured data for New Delhi.

	H (measured) (MJ/m^2)		*H* (estimated) (MJ/m^2)		ARE (%)	
Month	**Global**	**Diffuse**	**Global**	**Diffuse**	**Global**	**Diffuse**
January	12.32	5.21	12.90	4.89	2.2	6.2
February	16.42	6.22	15.80	5.82	2.8	6.5
March	20.64	7.56	13.71	7.96	4.5	5.3
April	24.07	3.83	22.18	3.39	2.7	5.0
May	24.43	3.68	22.67	3.16	2.1	4.9
June	22.54	3.66	22.20	12.16	2.9	4.3
July	13.07	3.83	13.20	3.01	4.6	6.9
August	17.79	3.27	16.86	3.53	5.2	7.2
September	13.90	3.27	13.95	7.64	5.6	7.6
October	16.80	6.37	16.07	5.92	4.3	7.0
November	14.13	4.92	12.73	4.61	2.8	6.2
December	3.93	4.87	3.56	5.18	2.1	6.4

Table 2.13 ARE percentages in the estimation of monthly mean solar irradiance in comparison with measured data for Jodhpur.

	H (measured) (MJ/m^2)		*H* (estimated) (MJ/m^2)		ARE (%)	
Month	**Global**	**Diffuse**	**Global**	**Diffuse**	**Global**	**Diffuse**
January	15.89	4.23	15.41	2.97	2.0	6.2
February	13.09	5.33	17.51	4.97	2.2	6.7
March	21.16	6.67	20.42	6.28	2.5	5.9
April	22.57	7.97	21.71	3.36	2.8	5.0
May	22.56	3.56	22.32	3.13	2.1	4.5
June	22.73	3.77	022.1	3.25	2.8	4.8
July	13.21	3.90	20.30	12.76	5.7	7.2
August	17.91	3.00	16.99	3.21	5.1	7.2
September	13.84	7.43	20.81	6.85	4.9	7.8
October	13.35	5.25	13.62	4.90	2.8	6.7
November	16.68	4.13	16.18	4.38	2.0	6.0
December	15.04	2.98	14.60	2.72	2.9	6.6

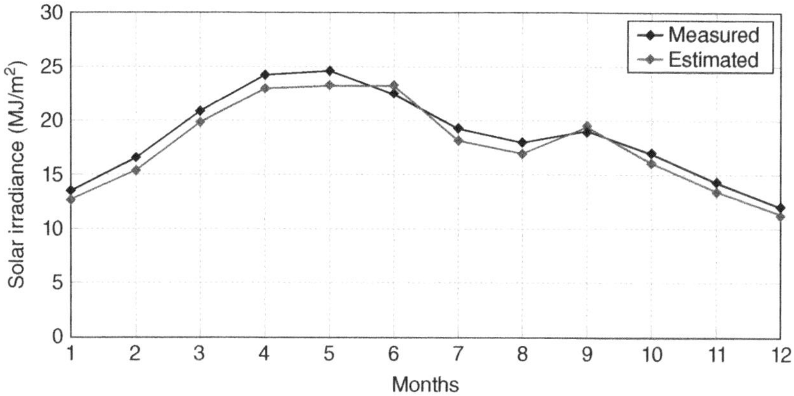

Figure 2.26 Measured and estimated global solar irradiance for New Delhi.

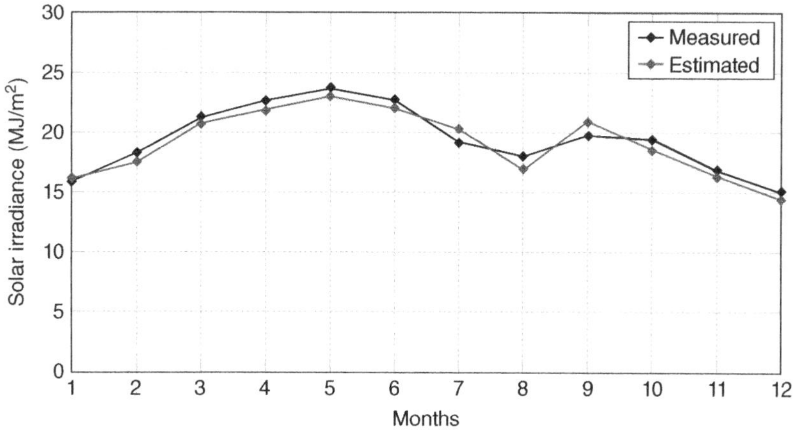

Figure 2.27 Measured and estimated global solar irradiance for Jodhpur.

Delhi, Jodhpur, Nagpur, Vishakhapatnam, Kolkata, Ahmedabad, and Shillong are shown in Figures 2.26 through Figures 2.32, respectively.

However, Figures 2.33 through 2.39 for New Delhi, Jodhpur, Nagpur, Vishakhapatnam, Kolkata, Ahmedabad, and Shillong, respectively, also graphically illustrate estimated values of diffuse solar irradiance using fuzzy models in comparison to the measured values.

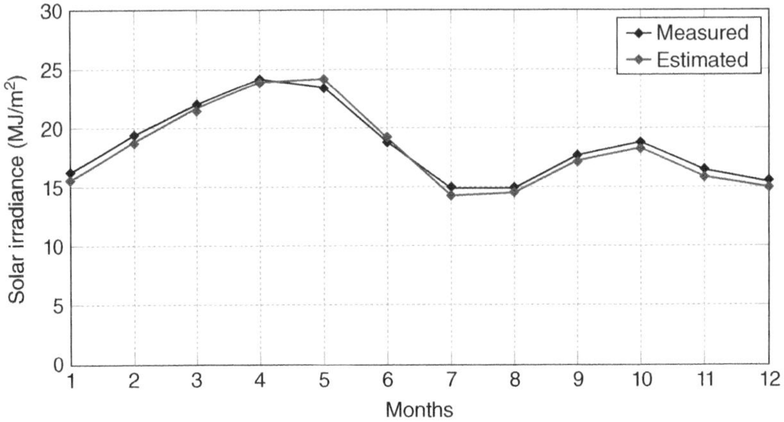

Figure 2.28 Measured and estimated global solar irradiance for Nagpur.

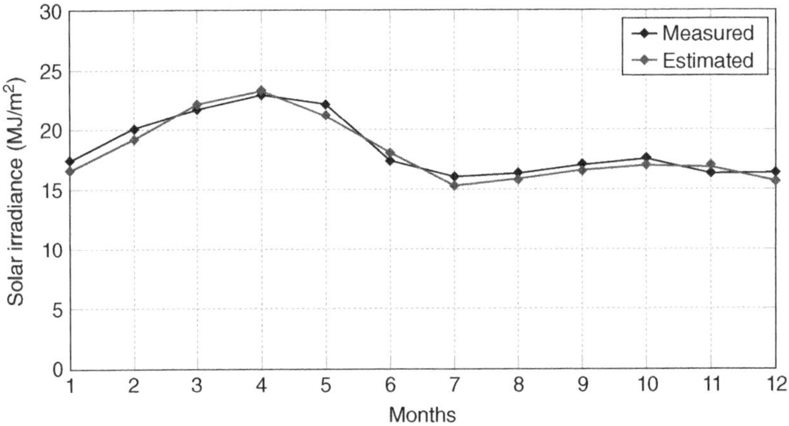

Figure 2.29 Measured and estimated global solar irradiance for Vishakhapatnam.

2.16 Artificial Neural Network for Solar Energy Estimation

The fundamental building blocks of the human brain are neurons, which provide us the capacity to integrate past experiences into the present. ANNs are computing algorithms that imitate the four fundamental tasks carried out by biological neurons: receiving inputs from other neurons or sources, combining them, performing operations on the result, and producing the finished product. ANNs are fascinating because, once they are set up, they can learn in a self-organizing manner that

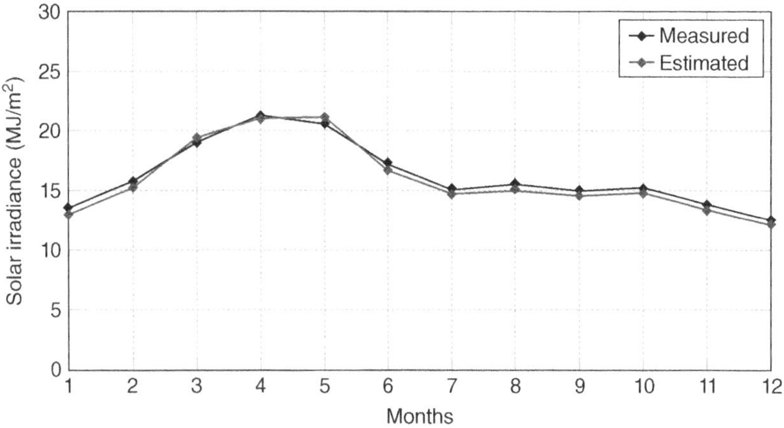

Figure 2.30 Measured and estimated global solar irradiance for Kolkata.

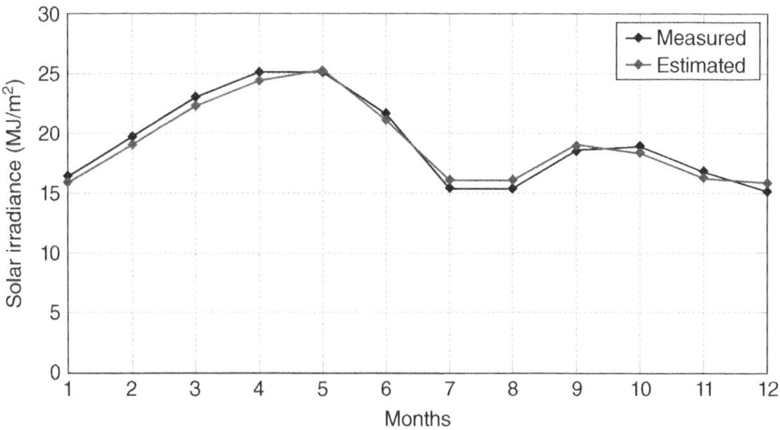

Figure 2.31 Measured and estimated global solar irradiance for Ahmedabad.

mimics brain processes like pattern recognition, classification, and optimization. ANNs can manage huge and complex.

The numerous interconnected characteristics focus on the more significant inputs and ignore extraneous material that is less vital. The back-propagation method and its variations are the most well-known and effective learning technique in neural networks. The mistake corrective learning rule is the foundation of this method. The error-correcting back-propagation procedure essentially entails two trips through the various network layers: a forward pass and a backward pass. An activity pattern is applied to the network's sensory nodes during

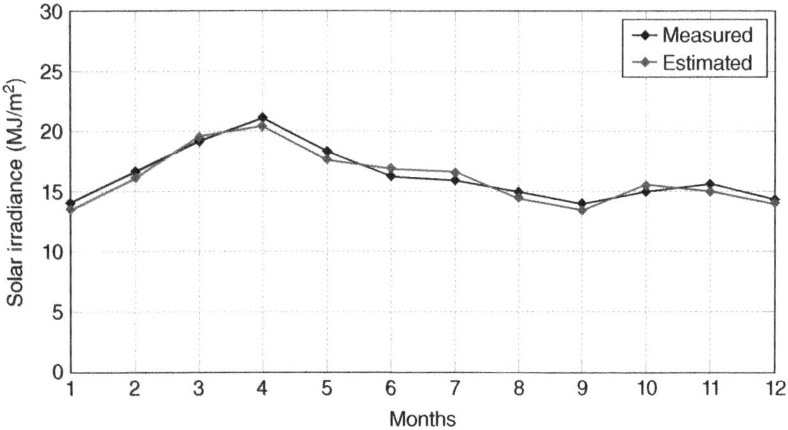

Figure 2.32 Measured and estimated global solar irradiance for Shillong.

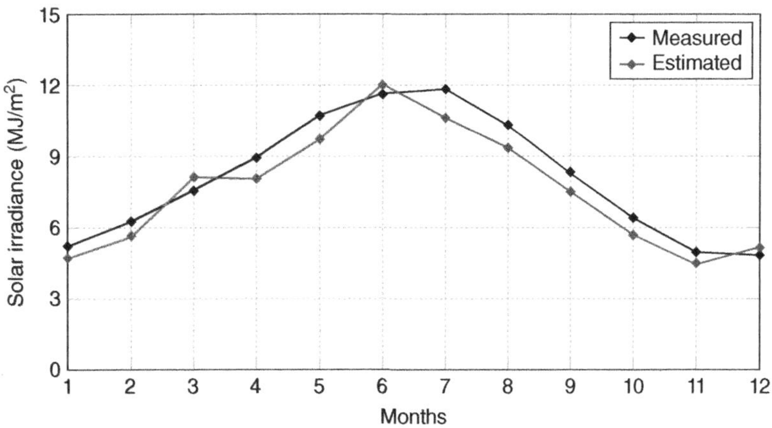

Figure 2.33 Measured and estimated diffuse solar irradiance for New Delhi.

the forward pass, and its effect spreads through the network layer by layer. As the network's real reaction, a collection of outputs is finally generated. On the other hand, during the backward pass, the network's synaptic weights are all changed in accordance with the mistake correction rule. To produce an error signal, the network's actual response is specifically deducted from a desired answer. The erroneous signal is then transmitted via the network in reverse.

The architecture, learning or training algorithm, and activation function of an ANN define it. The architecture explains how the neurons are connected. As

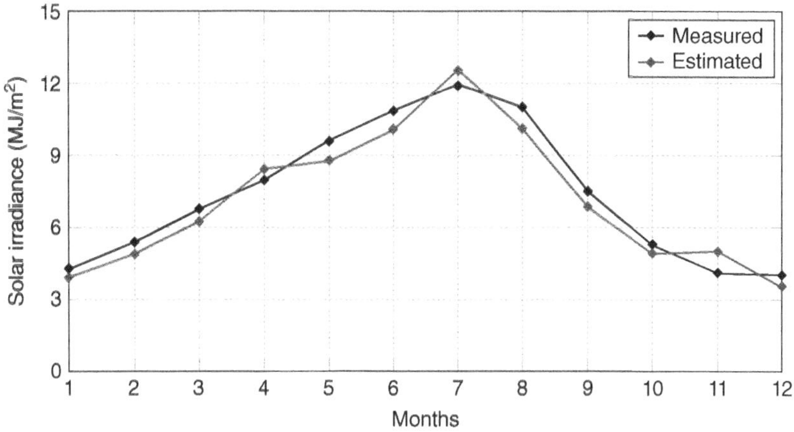

Figure 2.34 Measured and estimated diffuse solar irradiance for Jodhpur.

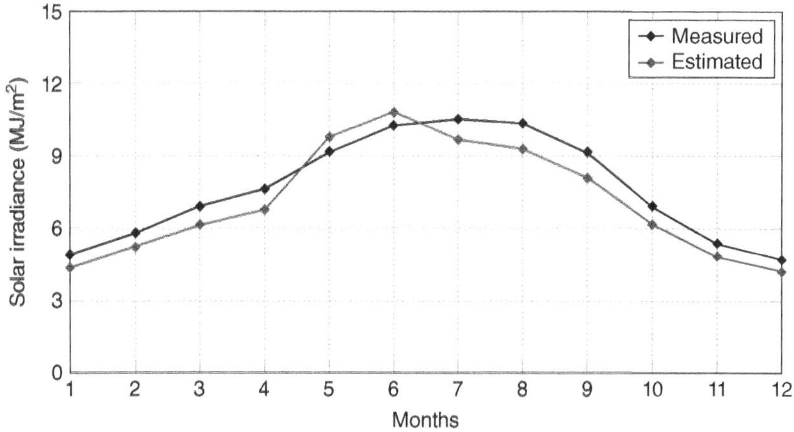

Figure 2.35 Measured and estimated diffuse solar irradiance for Nagpur.

shown in Figure 2.40b and c, it consists of an input layer, an output layer, and typically one or more hidden layers.

2.16.1 Artificial Neuron Model

Figure 2.40a depicts the artificial neuron model. In this paradigm, the processing element (neuron) determines whether the weighted sum of its inputs and outputs is higher or lower than a predetermined threshold, k. The externally applied bias

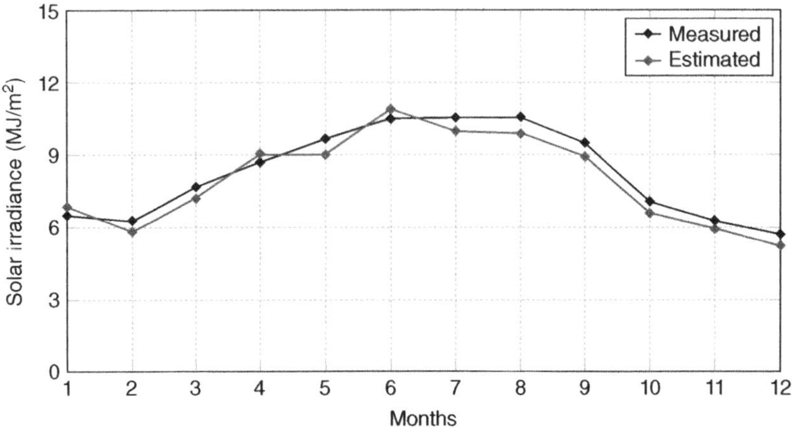

Figure 2.36 Measured and estimated diffuse solar irradiance for Vishakhapatnam.

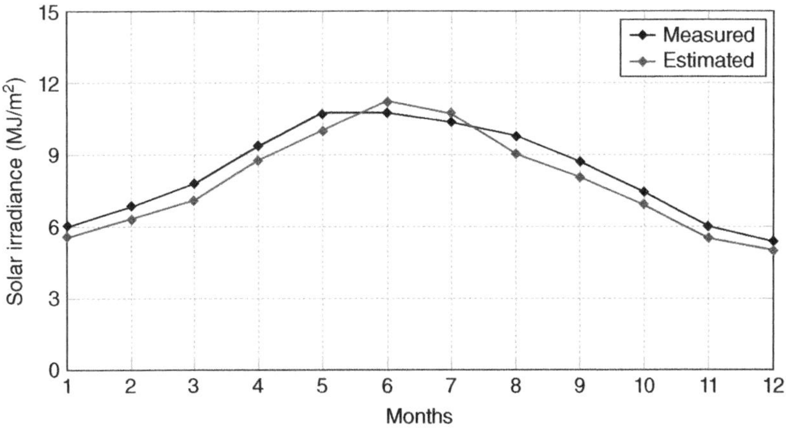

Figure 2.37 Measured and estimated diffuse solar irradiance for Kolkata.

has the effect of reducing the activation function's net input:

$$y_k = f(u_k \theta_k) \tag{2.60}$$

$$u_k = \sum w_{kj} x_i \tag{2.61}$$

Here, $x_1, x_2, \ldots, x_p$ are input signals, while $w_{k1}, w_{k2}, \ldots, w_{kp}$ are the neuron k's connectivity weights. U_k is the output that has been linearly combined.

A multilayer feed-forward network is created when various layers are connected, as seen in Figure 2.40c. A hidden layer sits between the input and output layers and

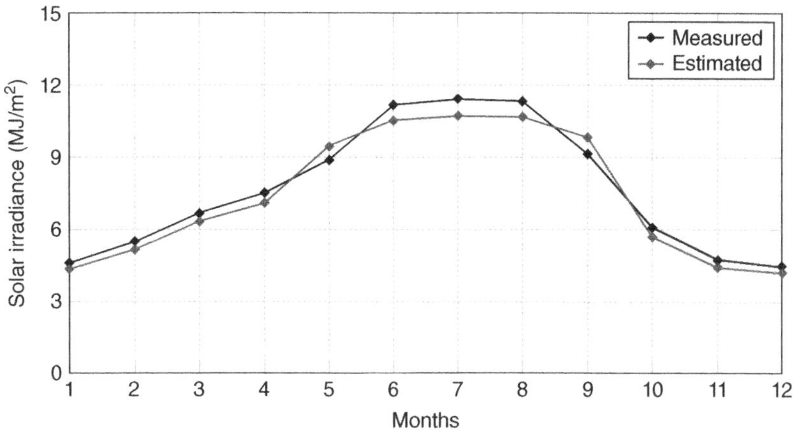

Figure 2.38 Measured and estimated diffuse solar irradiance for Ahmedabad.

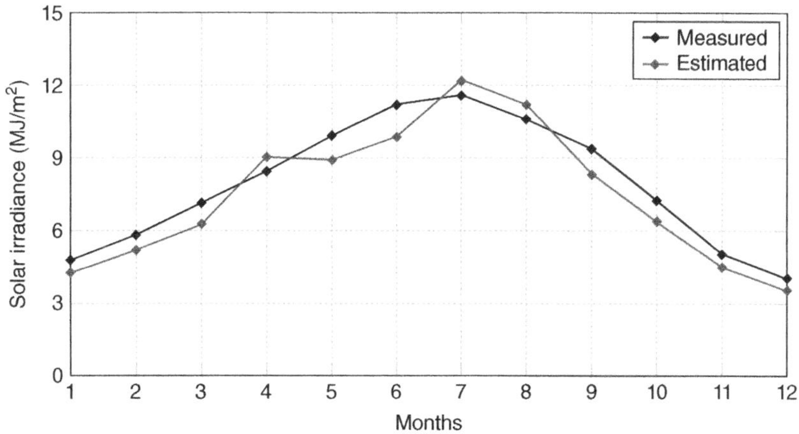

Figure 2.39 Measured and estimated diffuse solar irradiance for Shillong.

serves as an intermediary between external input and output. There was no direct interaction between the concealed layer and the outside world. A unique class of multilayer feed-forward networks includes the radial basis function network. A radial basis function, such as a Gaussian kernel, is used in this case as the activation function for the hidden layer. Recurrent networks are feedback networks with closed loops. The output of a processing unit in a single-layer recurrent network is feedback to that element, to another processing element, or to both. The neuron output in the multilayer recurrent network can be sent back to the nodes in the lower layers.

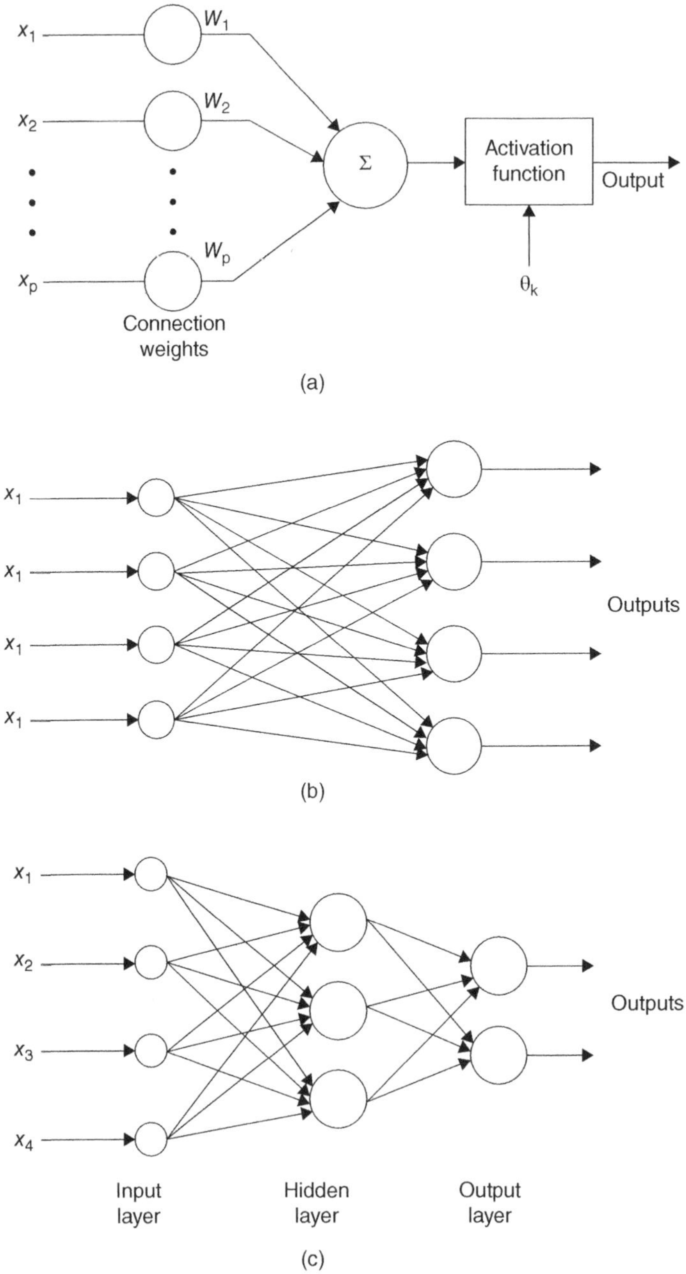

Figure 2.40 (a) Artificial neuron model. (b) Single-layer feed-forward network. (c) Multilayer feed-forward network.

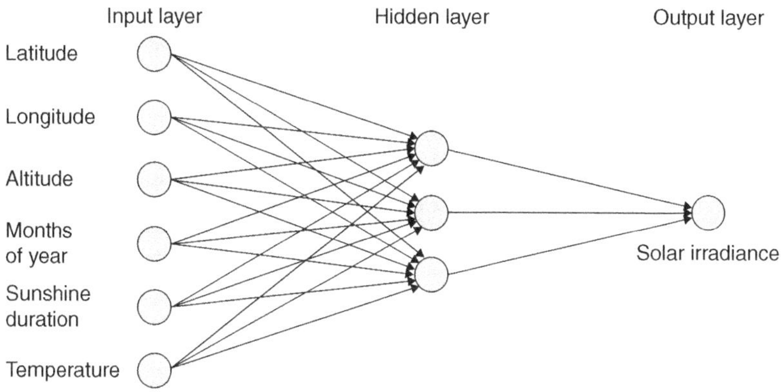

Figure 2.41 Proposed ANN architecture.

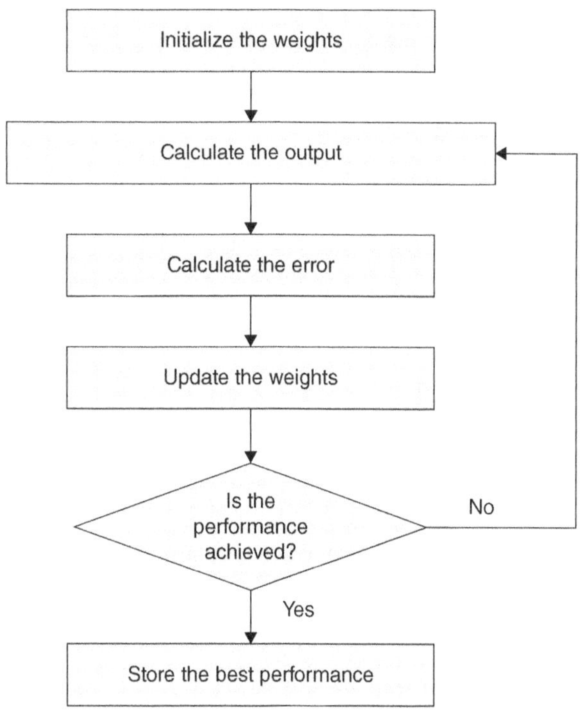

Figure 2.42 Flow chart for the proposed ANN model.

The architecture utilized in this model, which is depicted in Figure 2.41, consists of a hidden layer with a tan-sigmoid activation function, denoted by o_j, and an input layer with six inputs.

$$o_j = 1/(1 + e^{-n}) \tag{2.62}$$

where the appropriate input is n. The feed-forward back-propagation network is implemented using the MATLAB neural network toolkit. Figure 2.42 displays the flow chart for the same algorithm.

2.16.2 Normalization of Meteorological Data

If actual monthly data are used, the neural network's input and output data ranges might be different. Convergence issues could arise as a result throughout the learning process. The input and output data are scaled so that they stay within the range of 0.1–0.9 in order to avoid such issues. The lower extreme limit, which is 0, could not be exceeded during testing because the lower limit is 0.1. Similar to the lower limit, 0.9, the data is protected against possibly reach the 1.0 top extreme limit during testing. These 0.1-per-side margins are referred to as safe margins. The following equation for the solar energy estimating issue is used to scale the real data.

$$L_S = \frac{(Y_{max} - Y_{min})}{(L_{max} - L_{min})}(L - L_{min}) + Y_{min} \tag{2.63}$$

where

L = actual information
L_s = the scaled data that the network uses as input
L_{max} = the highest value in a specific dataset
L_{min} = the dataset's minimum value
Y_{max} = upper bound of normalizing range (0.9)
Y_{min} = lower bound of the normalizing range (0.1)

The air temperature, diffuse irradiance, global irradiance, and sunshine duration are the data that were used to create the proposed work. By dividing them by an alien value, these data are normalized. Seven Indian stations are chosen for the analysis: New Delhi, Kolkata, Mumbai, Jodhpur, Pune, Ahmedabad, Vishakhapatnam, and Shillong. These stations will be used to determine the monthly mean global and diffuse solar irradiance based on the length of sunshine and temperature.

2.16.3 Drawbacks of Conventional ANN

The traditional neural network model has the following significant flaws:

1) The typical neural network requires too much training time, which slows the system's reaction.
2) It is impossible to estimate how many hidden layers and hidden neurons there are, but they are significant in terms of approximating complex functions.
3) The current neuron model merely summarizes the weighted inputs; it does not conduct the product operation on those inputs.
4) Both the training time and the accuracy of the test result are influenced by the threshold (activation) function.
5) Back-propagation learning also has significant drawbacks, such as slow learning and the potential for local minima in the system.
6) Training data are affected by the normalization range. Therefore, choosing an appropriate range (i.e. maximum and lowest values) is crucial since it has an impact on the outcomes of neural network training.
7) The neural network's training time is based on how the input–output pattern (I/O mapping) is provided to the network.
8) The order in which the data are given also affects how long the network takes to train.

Several variations have been created throughout the years to address these shortcomings. The majority of the alternatives either raise the processing workload or burden learning algorithms. In this thesis, the estimation of solar energy is addressed using a new generalized neuron model that addresses the shortcomings of the traditional neural network by conducting a number of different alterations and adjustments on the prior model. The system's nonlinearities should be incorporated into the model. The following characteristics ought to be included in the model:

1) Both simple and high-order neuronal features should be present in the generalized neural network.
2) There is no requirement to choose the number of neurons and hidden layers, which should lower the network's complexity.
3) The network's response shouldn't be impacted by the input–output mapping.
4) There shouldn't be a normalizing impact.

2.17 Conclusion

Accurately measuring and estimating solar irradiance is crucial for the optimal performance of solar energy systems and the advancement of renewable energy technologies. The integration of reliable irradiance data allows for enhanced energy production forecasting, efficient PV system design, and effective grid management. Various methodologies, including ground-based measurements using pyranometers and pyrheliometers, satellite-derived estimates, and machine learning techniques, have proven effective in providing high-resolution and real-time irradiance data. Each method comes with trade-offs in terms of accuracy, cost, and ease of deployment.

While direct measurements offer high precision, they can be limited by geographic coverage and maintenance requirements. On the other hand, satellite-based and model-driven estimates offer wider coverage but may involve uncertainties due to atmospheric conditions and spatial resolution limits. Emerging technologies like artificial intelligence and data fusion techniques hold significant promise for bridging these gaps, providing more accurate, reliable, and real-time solar irradiance predictions.

In conclusion, continuous advancements in solar irradiance measurement and estimation are vital for the scalability of solar power systems. By leveraging innovative estimation models and refining measurement instruments, the global transition toward cleaner energy can be accelerated, contributing to more sustainable and resilient energy infrastructures.

References

1 Pereira, S., Canhoto, P., and Salgado, R. (2024). Development and assessment of artificial neural network models for direct normal solar irradiance forecasting using operational numerical weather prediction data. *Energy and AI* 15: 100314.

2 Zhu, H., Jiang, T., Zhou, H. et al. (2024). A novel soft sensing method using intelligent modeling method for solar irradiance and temperature in distributed PV power plant. *Energy Science & Engineering* 12 (3): 1041–1058.

3 Esposito, E., Leanza, G., and Di Francia, G. (2024). Comparative analysis of ground-based solar irradiance measurements and Copernicus satellite observations. *Energies* 17 (7): 1579.

4 Tercha, W., Tadjer, S.A., Chekired, F., and Canale, L. (2024). Machine learning-based forecasting of temperature and solar irradiance for photovoltaic systems. *Energies* 17 (5): 1124.

5 Akhund, T.M.N.U., Nice, N.T., Joy, M.A. et al. (2024). Anomaly prediction in solar photovoltaic (PV) systems via Rayleigh distribution with integrated internet of sensing things (IoST) monitoring and dynamic sun-tracking. *Information* 15 (8): 451.

6 Xiu-Yan, G., Jie-Mei, L., Yuan, Y., and He-Ping, T. (2024). Global horizontal irradiance prediction model considering the effect of aerosol optical depth based on the Informer model. *Renewable Energy* 220: 119671.

7 Zhang, S., Wang, J., Ji, X. et al. (2024). Oxygen vacancies modulating performance for Ga_2O_3 solar-blind photodetectors via low-cost mist chemical vapor deposition. *Materials Today Communications* 39: 108717.

8 Sebastianelli, A., Serva, F., Ceschini, A. et al. (2024). Machine learning forecast of surface solar irradiance from meteo satellite data. *Remote Sensing of Environment* 315: 114431.

9 İlhan, C. and Çalık, Z. (2024). Solar-powered UAV: a novel approach to conceptual design. *Konya Journal of Engineering Sciences* 12 (2): 396–409.

10 Jia, J., Menenti, M., Jia, L. et al. (2024). Estimation of all-sky solar irradiance components over rugged terrain using satellite and reanalysis data: The Tibetan Plateau Experiment. *IEEE Transactions on Geoscience and Remote Sensing* 62: 1–23.

11 Thaker, J., Höller, R., and Kapasi, M. (2024). Short-term solar irradiance prediction with a hybrid ensemble model using EUMETSAT satellite images. *Energies* 17 (2): 329.

12 Riise, H.N., Nygård, M.M., Aarseth, B.L. et al. (2024). Benchmark of estimated solar irradiance data at high latitude locations. *Solar Energy* 282: 112975.

13 AlKandari, M. and Ahmad, I. (2024). Solar power generation forecasting using ensemble approach based on deep learning and statistical methods. *Applied Computing and Informatics* 20 (3/4): 231–250.

14 Sammar, M.J., Saeed, M.A., Mohsin, S.M. et al. (2024). Illuminating the future: a comprehensive review of AI-based solar irradiance prediction models. *IEEE Access* 12: 114394–114415.

15 Saini, K.K., Sharma, P., Mathur, H.D. et al. (2024). Techno-economic and reliability assessment of an off-grid solar-powered energy system. *Applied Energy* 371: 123579.

16 Al-Dahidi, S., Madhiarasan, M., Al-Ghussain, L. et al. (2024). Forecasting solar photovoltaic power production: a comprehensive review and innovative data-driven modeling framework. *Energies* 17 (16): 4145.

17 Almarshoud, A. (2024). Validation of satellite-derived solar irradiance datasets: a case study in Saudi Arabia. *Future Sustainability* 2 (2): 1–7.

18 Hendrikx, N.Y., Barhmi, K., Visser, L.R. et al. (2024). All sky imaging-based short-term solar irradiance forecasting with Long Short-Term Memory networks. *Solar Energy* 272: 112463.

19 Chaaban, A.K. and Alfadl, N. (2024). A comparative study of machine learning approaches for an accurate predictive modeling of solar energy generation. *Energy Reports* 12: 1293–1302.

20 Bouquet, P., Jackson, I., Nick, M., and Kaboli, A. (2024). AI-based forecasting for optimised solar energy management and smart grid efficiency. *International Journal of Production Research* 62 (13): 4623–4644.

21 Tan, L., Kang, R., Xia, J., and Wang, Y. (2024). Application of multi-source data fusion on intelligent prediction of photovoltaic power. *Solar Energy* 277: 112706.

22 Han, T., Li, R., Wang, X. et al. (2024). Intra-hour solar irradiance forecasting using topology data analysis and physics-driven deep learning. *Renewable Energy* 224: 120138.

3

Parabolic-Trough Concentrating Solar Power (CSP) Systems

3.1 Introduction

A parabolic-trough concentrator (PTC) is a type of linear-focus solar collector that reflects direct solar energy onto a receiver or absorber tube that is situated in the parabola's focal line (see Figure 3.1). The bigger collector aperture area concentrates direct solar energy that has been reflected onto the smaller receiver tube's outer surface, heating the fluid that flows through it. Thus, the solar radiation is converted into thermal energy in the form of the fluid's sensible or latent heat.

This thermal energy can then be utilized to fuel either Rankine cycles to generate electricity using a steam turbine in a "solar thermal" power plant, or industrial activities that require thermal energy (such as the food industry and the petrochemical industry). Parabolic troughs can generate useful thermal energy up to 398°C with today's technology. Due to the thermal oil's rapid degradation over 398°C, which is currently employed as the working fluid, the maximum temperature is primarily constrained by this factor. In the medium future, higher temperatures approaching 500°C are anticipated thanks to research into novel fluids.

Historical development era BC. According to Diocles, a parabolic reflector directs solar rays toward a common point that is situated at a certain location (i.e. the parabola's focal point). However, it took more than 20 centuries for the first solar system to actually be built utilizing parabolic trough collectors. John Ericsson, a Swedish engineer, created the first parabolic-trough solar collector that was vividly depicted around the end of the nineteenth century. This collector generated saturated steam for a little 373 W steam engine with a solar radiation collecting surface of 3.25 m^2. The parabolic shape with a 3.35, 4.9 m aperture was added by Ericsson in 1883 by installing flat strips of silvered-glass reflectors in a frame [1]. This enhanced the original design, which used reflectors composed of polished sheet metal. With a single 0.153 m diameter 0.2 m stroke piston, it produced saturated steam at a pressure of 0.24 MPa to power a steam engine

Concentrated Solar Power Systems, First Edition. Bellamkonda Pragathi and D. P. Kothari.

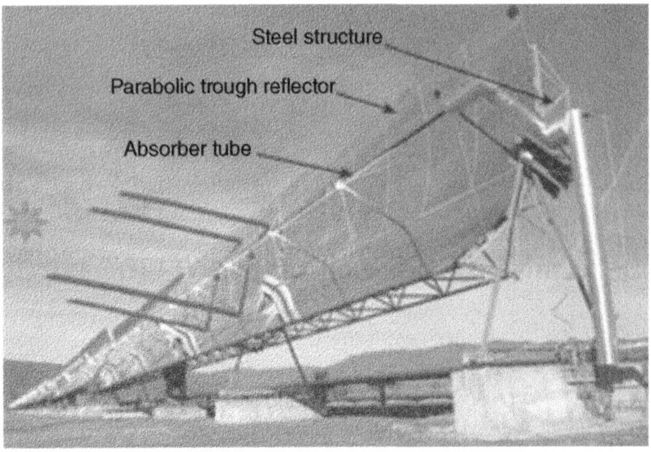

Figure 3.1 The Greek mathematician Diocles wrote the first text detailing the optical characteristics of a parabolic-trough collector in the second century BC.

operating at 120 rpm. Because Ericsson always kept the technical specifics a secret, any chance of commercial development died with him in 1883 [2]. Frank Shuman, an American engineer, created a parabolic-trough collector solar plant in Egypt in 1913 that produced 0.1 MPa saturated steam to power a steam engine that pumped irrigation water for a farming settlement in Meadi, close to Cairo. Shuman modified the overall plant design with the help of British professor C.V. Boys and the financial backing of British investors. One of the five collectors put in at Meadi in 1913 is shown in Figure 3.2. Each collector had a 4.2 m wide parabola and a length of 62.17 m, adding up to a solar field collecting surface of 1274.5 m^2 with a 40% collector efficiency. With the help of the solar field's energy, 380 l/s of water was pumped. A metal framework on a concrete foundation supported the reflectors and reception tubes. The image in Figure 3.2 was taken from a Hans Gunther book published in 1922 [3].

The similarities between the parabolic-trough collectors created by Shuman in 1912 and the collectors used in solar thermal power plants today are evident in the photograph and the digitalization needed for reprinting here. However, no additional similar facilities were constructed at that time due to World War I and the advent of oil into the energy market [4]. Twenty years later, in 1935, C.G. Abbot used a PTC and a 0.37 kW steam engine to transform solar energy into mechanical power. After that, there was no more notable PTC development in the twentieth century until Sandia National Laboratories, Honeywell International Inc., and Westinghouse developed and tested three PTC prototypes in the United States and conducted a thorough cost analysis in 1979, which coincided with a resurgence in interest in solar energy following the 1973 oil price crisis .Several other businesses created new PTC designs a short while later, in the 1980s, and joined the market

Figure 3.2 One of the five parabolic-trough collectors installed by Frank Shuman at Meadi (Egypt) in 1913. Source: Moya [5]/Elsevier.

with scaled-down industrial process heat applications and miniaturized solar thermal power plants. Table 3.1 covers four modest demonstration solar thermal power plants that were constructed at the time in the United States, Japan, Spain, and Australia. In the early 1980s, several companies, including General Electric, Honeywell Inc., Acurex Solar Corp. Solar Kinetics Corp. and Jacobs Del. Corp., manufactured and sold various PTCs. At the same time, the Helioman 3/32, the only two-axis PTC ever marketed, was created by the German business Maschinenfabrik Augsburg-Nürnberg (MAN) [6].

In spite of the fact that later on, other businesses—among them the American business Industrial Solar Technologies (IST)—entered the market with successful PTC designs; they tended to be more for industrial process heat applications. The design and deployment of the nine Solar Electricity Generating System (SEGS) plants in the Mohave Desert (California, USA) by LUZ International

Table 3.1 Details of demonstration trough-based solar thermal power plants built during the early 1980s.

	Coolidge (United States)	**Sunshine (Japan)**	**IEA-DCS (Spain)**	**STEP-100 (Australia)**
Net electric power	0.15 MW_e	1 MW_e	0.5 MW_e	0.1 MW_e
Total aperture area	2,140 m^2	12,856 m^2	7,622 m^2	920 m^2
Heat transfer fluid	Synthetic oil	Water/steam	Synthetic oil	Synthetic oil
Effective storage capacity	5 MW_{th}	3 MW_e	0.8 MW_e	117 MW_{th}
Duration of service	1980–1982	1981–1984	1981–1987	1982–1988

Figure 3.3 View of one of the SEGS plants.

Limited from 1984 (SEGS-I) through 1990 (SEGS-IX) was the most notable PTC technology-related event of the twentieth century. One of the nine SEGS plants is shown in Figure 3.3, and the operating principle is schematically shown in Figure 3.4 [7]. Through the receiver pipes of the trough array, a synthetic heat transfer fluid (HTF) is pumped and heated by concentrated solar light. The steam created by these heat exchangers is then cycled back to the solar field using this oil. A typical steam turbine-based energy producing facility uses the steam [8]. The SEGSs generally rely on natural gas burning to provide continuous operation when the sun is not available, despite the fact that some hot oil-based energy storage was included in the initial plant [9]. The SEGS plants,

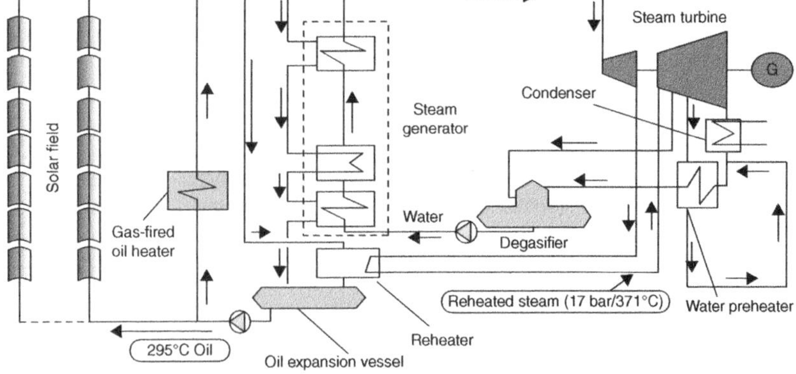

Figure 3.4 Configuration of a typical SEGS plant.

which have a combined electrical output of 354 MW and more than two million square meters of parabolic-trough collectors, have been a crucial part of the advancement and commercial deployment of parabolic-trough collectors at the start of the twenty-first century [10]. For both investors and promoters, the fact that these plants are still in continuous use after all these years offers a high level of technical credibility and confidence. But in addition to serving as a "technology showcase" for new projects promoted in the twenty-first century, the SEGS plants have helped the commercial take-off of parabolic-troughs by enabling the existence of an industrial capability to at least partially satisfy the initial demand for essential components, such as receiver tubes and reflectors [11].

The advantageous conditions outlined by the legal framework put in place in the United States as a result of the sharp increase in oil prices in the 1970s led to the development of the SEGS facilities. However, when the price of oil plummeted once more in the 1980s, these government incentives for renewable systems were reduced and were insufficient, making the installation of more SEGS units impractical. Government incentives have once more been put in place in several nations since the turn of the century, boosting a wide range of solar thermal power plant projects (e.g. tax credits in the United States and a favorable feed-in tariff in Spain and Algeria) [12]. Due to the SEGS plants' long-term track records of performance, the majority of these new projects are based on parabolic-trough collectors because they are the "least risky" and easiest to finance option.

An increase in the number of solar power plants has resulted in major R&D spending, the creation of numerous new parabolic-trough collector designs, and the establishment of new factories in the United States and Spain for manufacturing vital parts like reflectors and receiver tubes.

3.2 Commercially Available Parabolic-Trough Collectors (PTCs)

3.2.1 Large PTCs

One of LUZ International's accomplishments was the creation of three dependable, long-lasting PTC designs, known as the LS1, LS2, and LS3, which were successfully installed and operated in the SEGS plants. Larger PTCs would need to be created for larger CSP plants, even though the LS1 installed by LUZ at the SEGS-I plant in 1984 had a unit length of 50.2 m and a parabola width of 2.5 m, which was comparable in size to other designs created during the early 1980s [13]. Due to this, LUZ created the LS2 and LS3 designs, having 235 and 545 m^2 of aperture space per collector, respectively. Lack of an appropriate PTC was one of the obstacles to the construction of huge solar fields using parabolic troughs after

Figure 3.5 The steel structure of the Eurotrough-100 collector design. Source: Lüpfert et al. [14]/The American Society of Mechanical Engineers.

LUZ's demise. In 1998, a European partnership made up of business, engineering companies, and R&D facilities was established to create a new PTC appropriate for big CSP plants [15]. The outcome was the creation of the Eurotrough-100 (ET-100) and Eurotrough-150 (ET-150) designs, which were then enhanced to produce its successor, the SKAL-ET, the collector that was set up in 2007 at the ANDASOL-I plant in Spain (Figure 3.5; Tables 3.2 and 3.3) [16].

To obtain the ideal parabolic shape of the concentrator and overall structural rigidity, while also keeping the cost of assembly low, highly precise assembly of the steel structural profiles is necessary. As a result, even though not all PTC systems need assembly jigs, the majority of contemporary designs must adhere to the design tolerances, which are typically 1 mm. A final inspection of an assembly jig for the Eurotrough collector design is shown in Figure 3.6.

To provide good PTC optical and geometrical performance under moderate wind speeds (below 35 km/h), the collector design in Figure 3.7 features a torque box, a central space frame that provides good rigidity and resists twisting. Figure 3.6 also depicts the final appearance of a Eurotrough parabolic-trough module after all of its parts have been installed in the assembly jig. Two or three assembly lines are employed in parallel in big parabolic-trough solar fields to speed up construction. A different PTC design strategy involves swapping out the torque box with a central steel tube known as the torque tube. However, it's also crucial to assemble the steel mirror support frames on this central tube precisely.

Table 3.2 Parameters of the Eurotrough-150 collector design.

Parabola width (m)	**5.76**
Overall length of a single collector (m)	143.5
Length of every module (m)	12.27
Number of parabolic trough modules	12
Outer diameter of the steel receiver pipe (m)	0.07
Inner diameter of the steel receiver pipe (m)	0.065
Collector aperture area (m^2)	822.5
Mirrors reflectivity	0.93
Steel receiver pipe absorptance	0.95
Intercept factor	0.90
Receiver pipe glass cover transmittance	0.95
Peak optical efficiency	0.75

Table 3.3 The specifications for the ET-150 collector design and the steel frame of the ET-100 design.

	IST-RMT	**IST-PT1**	**PCT-1800**	**Soponova 4.0**
Structure	Steel/ aluminum	Steel/ aluminum	Aluminum	Steel/ aluminum
Reflectors	Polished aluminum	Aluminum acrylic	Al-coating 0.5 mm	Aluminum
Trough module size	3.66 × 1.13 m	6.1 × 2.9 m	5.09 × 1.8 m	3.66 × 1.52 m
Receiver tube	Nonevacuated, with black chrome coating	Nonevacuated, with black chrome coating	Nonevacuated, with selective coating	Nonevacuated, with selective coating
Maximum temperature	204°C	288°C	~200°C	260°C

Figure 3.7b is a close-up of the torque tube-based collector created by the Spanish business Albiasa Solar. In addition, torque tubes rather than torque boxes are used in the collectors created by the Spanish firms SENER (www.sener.es) and URSSA (www.urssa.es) between 2006 and 2013. PTC torque box designs typically perform better and are more rigid under wind loads, but their main drawback is a greater assembly cost caused by the sheer volume of steel profiles that must be assembled

(a)

(b)

Figure 3.6 Check out of the assembly jig for Eurotrough collectors (a) and a parabolic-trough module assembled in the jig (b). Source: Moya [5]/Elsevier.

with extreme precision. On the other hand, torque tube designs are often a bit more affordable but are more susceptible to distortion than those who use a torque box, both bending and torsion caused by gravity and wind loads. In any case, the rigidity of both PTC designs under wind loads is sufficient to prevent deformation from exceeding acceptable bounds. There are also commercial PTC designs that offer a strong structure without the use of a torque box or torque tube and instead use a metallic space frame. Examples of these designs include the 430.8 m^2 collector aperture area used by Solargenix and Acciona in the Nevada Solar One Plant and the 690 m^2 SkyTrough collector from the American company. The design employed by Solargenix and Acciona is noteworthy in that it is one of the few PTC designs that depends on the accuracy of the preformed pieces and

(a)

(b)

Figure 3.7 PTC designs using a space frame (a) and torque tube (b) to provide good torsional rigidity. Source: Moya [5]/Elsevier.

fasteners to yield an accurate shape rather than using an accurate jig for assembly. The space frame, which exhibits good torsional rigidity and performance in wind loads, was created by the company Gossamer Space Frames for Solargenix and successfully employed in the CSP facility Nevada Solar One.

There are other methods for generating a stiff structure that can lower costs while yet maintaining high stiffness and assembly precision. However, at the end of 2010, these groundbreaking concepts were still just in the prototype stage. An efficient quality control procedure is crucial during the entire assembly process to guarantee that design tolerances and performance are met because after the structures have been installed in the solar field, the size and number of components in the structure of large PTC collectors used in solar thermal power plants make correction of errors almost impossible. To prevent solar tracking mistakes, especially when open-loop tracking systems based on mathematical calculation of the sun coordinates are utilized, it is necessary to assemble the steel structures with extreme precision as well as to precisely position the parallel rows of collectors in the solar field. Parabolic-trough concentrating solar power (CSP) systems: Any misalignment would result in a tracking error, which would reduce the amount of focused solar radiation that reaches the receiver pipes. Although the curved back-silvered low-iron thick-glass mirror is the most common type of reflector used in commercial PTC designs, it can alternatively be built of polished sheet metal or silver- or aluminum-coated films that can be laminated onto a rigid parabola-shaped substrate. These polymer film reflectors are available from several providers. Examples of polymer film reflectors for solar applications include VikuitiTM, marketed by 3M, and ReflecTech®, marketed by Reflectech Inc. However, there are still little and fewer experimental data about their outdoor performance and durability than there are for glass mirrors.

3.2.2 Small PTCs

Although giant parabolic-trough collectors for large solar thermal power plants received the majority of R&D attention in the late twentieth and early twenty-first centuries, new, smaller troughs have also been created for process heat applications with temperatures below 300°C. These collectors can replace natural gas in various applications, such as crop drying and food preparation, by providing process heat. These tiny parabolic-trough collectors are capable of being used in industrial operations including biofuel generation, water purification, desalination, and absorption-chiller air conditioning for commercial and industrial buildings. In the United States, IST played a significant role in the development of numerous collectors and the installation of numerous commercial plants. When the Spanish industrial conglomerate Abengoa acquired IST in 2007, its IST-PT1 and IST-RMT collectors were marketed by Abengoa Solar IST. Both collectors have nonevacuated receiver pipes with a glass envelope and a black-chrome finish, and their respective maximum operating temperatures are 288 and 204°C. The PTC-1800 collector, invented by the German-Turkish company SOLITEM, is another tiny PTC for process heat applications. This collector's maximum operating temperature is 200°C, and its collecting aperture area is 36.65 m^2. A small PTC, the PolyTrough-1200, designed by the Australian business NEP Solar Pty Ltd, is likewise suited for temperatures up to 230°C. The composite

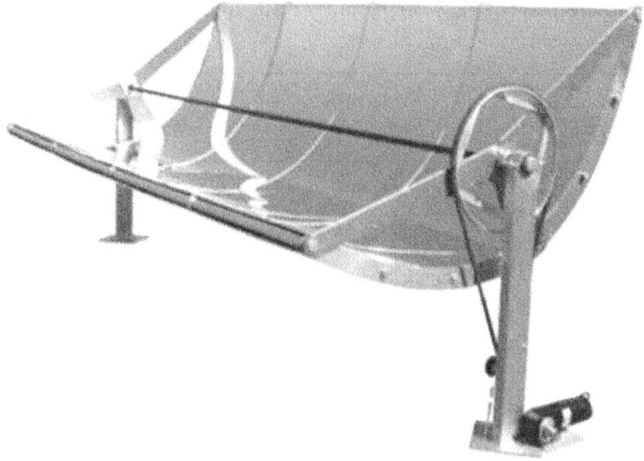

Figure 3.8 The Soponova 4.0 parabolic-trough concentrator developed by SOPOGY.

reflector panels with aluminum sheets that make up the 24 m long by 1.2 m wide PolyTrough-1200 collector can be placed on roofs or at the ground level. A 3.7 m long by 1.35 m wide parabolic-trough collector module known as Soponova 4.0 has been created by SOPOGY, a company with headquarters in Honolulu, Hawaii, and is appropriate for both process heat applications and the production of electricity. A Soponova 4.0 collector module is depicted in Figure 3.8. SOPOGY constructed eight solar power plants in various locations worldwide in 2013. The most current one opened in December 2009 with a 2 MW_e-rated power production in Keahole (Hawaii). Based on Soponova 4.0 trough-collector arrays, SOPOGY created a modest solar thermal power plant idea that is expandable from 250 kW_e to over 20 MW_e. SOPOGY filed a patent for MicroCSPTM, which stands for scalable solar thermal power plant. The technical specifications of the small parabolic-trough collectors sold by Abengoa Solar IST, SOLITEM, and SOPOGY.

3.2.3 Receivers

In reality, the conventional PTC receiver tube is made of two concentric pipes: an outer glass tube encircling the inner steel pipe, which houses the working fluid. To increase its transmittance of solar radiation, the glass tube is manufactured of low-iron borosilicate glass. The steel pipe's outer surface is an optically selective surface with a high solar absorptance and a low emittance of infrared light that is produced thermally. In Chapter 15, the fundamentals of these surfaces are covered in great detail. An antireflective coating is typically applied to the

Table 3.4 Technical parameters of the receivers commercialized by Schott, Siemens, and ASE.

	Schott PTR-70	**Siemens UVAC-2010**	**ASE HEMS08**
Solar absorptance	≥0.95	≥0.96	≥0.95
Solar transmittance	≥0.96	≥0.96	n.a.
Thermal emittance	≤0.1 at 400°C	≤0.09 at 400°C	≤0.1 at 400°C ≤0.14 at 580°C
Steel pipe inner/outer diameters	70/65 mm stainless steel	70/65 mm stainless steel	70/65 mm stainless steel
Thermal losses	250 W/m at 400°C	n.a.	230 W/m at 400°C
Glass cover	Borosilicate	Borosilicate	Borosilicate
Active length ratio at 350°C	>96%	96.4%	n.a.
Maximum fluid temperature	400°C	400°C	550°C

glass tube to increase solar transmittance and improve annual performance. There are two types of receivers for parabolic-trough collectors: evacuated and nonevacuated. Since they feature a strong vacuum (i.e. 10^5 mbar) between the steel pipe and the glass cover, evacuated receivers are frequently employed for temperatures exceeding 300°C because they minimize heat losses.

The typical evacuated receiver is shown in Figure 3.3. These receivers have stainless steel expansion bellows that connect the glass cover to the steel pipe. When the receiver tube is operating at nominal temperature, these bellows not only account for the different thermal expansion of glass and steel but also create a tight annular gap between the two tubes to create a vacuum. These expansion bellows have one end that is directly bonded to the steel pipe's outside, and the other end that is attached to the end of the glass cover via a glass-to-metal welding process. Chemical "getters," as seen in Figure 3.9, are positioned in the space between the glass cover and the steel receiver pipe to capture gas molecules that are moving from the fluid to the annulus through the steel pipe wall. Due to its technical complexity and high cost (about 850 C/unit in 2010), evacuated receivers are only used for higher temperatures where superior thermal efficiency is necessary and the high cost is offset by a higher thermal output. Only Schott, Siemens, and ASE were producing evacuated PTC receivers at the end of 2013. Prior to 2009, the majority of parabolic-trough solar thermal power plants installed worldwide used receivers made by either the German business Schott (www.schottsolar.com) or the Israeli company Solel (which Siemens acquired in 2009).

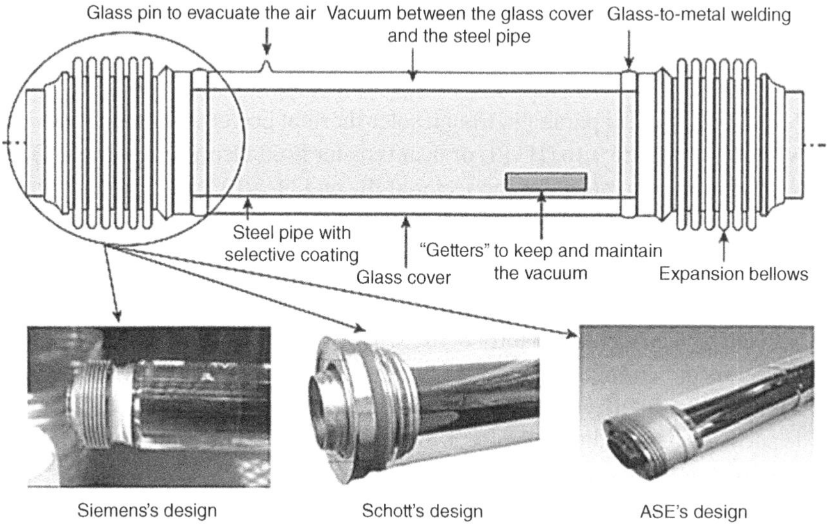

Figure 3.9 A typical evacuated receiver for parabolic-trough collectors. Source: Khan et al. [17]/Elsevier.

The Archimede Plant in Syracuse, Italy, was the first facility to use HEMS08 receivers. It began operations in 2010 and used molten salt as the receiver working fluid, which is a solution of sodium and potassium nitrate. Figure 3.3 demonstrates how these three manufacturers link the inner steel pipe and the glass cover using flexible bellows. In To event significant thermal and mechanical stress that could lead to the glass-to-metal welding used to join the glass cover to the flexible bellows, which is a weak point in the receiver tube, must be shielded from concentrated solar radiation. As a result, the flexible bellows are typically covered with an aluminum shield to safeguard the welding. The technical specifications of the receivers made by the Schott, Siemens, and ASE firms are displayed in Table 3.4. Because thermal losses are less important at working temperatures below 300°C, nonevacuated receivers are appropriate for these applications. Nonevacuated receivers similarly consist of an inner steel pipe and a glass cover, but they don't contain any vacuum or glass-to-metal welds in between the steel pipe and its glass cover. Selective coatings are less complex for nonevacuated receivers than for evacuated receivers. Coatings made of black-chrome or black nickel are frequently utilized since they are affordable and simple to make. They are connected in series up to the length of the PTC because the maximum receiver tube length is often less than 5 m due to manufacturing limitations. Nonevacuated receivers are often joined by specific threaded connections, while evacuated receivers are typically welded together.

3.3 Existing Parabolic-Trough Collector (PTC) Solar Thermal Power Plants

The layout of a typical parabolic-trough solar thermal power plant using thermal oil is shown in Figure 3.10 (HTF1, or heat transfer fluid technology). These facilities, which were the norm for the sector at the end of 2010, can be broken down into three subsystems:

1) **The solar system**: Between the solar field and the power block or thermal storage system, this is where direct solar radiation is gathered and transformed into thermal energy in the form of the sensible heat of the working fluid. Solar collectors are arranged in parallel rows throughout the solar field.
2) **The system for thermal energy storage (TES)**: This subsystem, which is located in the middle of Figure 3.10, is not necessary for the plant to function. Although a thermal storage system increases annual operating hours and, consequently, the amount of electricity produced, it also improves plant dispatchability and improves plant operation during cloud transients by sending all of the thermal energy generated by the solar field to the storage system rather than the power block. By acting as a thermal buffer between the steam and the steam turbine, this operating strategy prevents damage to the steam turbine from fluctuating steam characteristics.
   The fluid is referred to as a HTF because it transports heat from the solar field to the point of use and often differs from the working fluid of the power cycle (for power generation).
   Any temperature-related instability in the solar field exit is filtered by the power block's thermal inertia and the solar field. However, even without thermal storage, a big solar thermal power plant's steam generator and the enormous thermal inertia they give can still supply enough thermal energy to keep the power block operating at a constant output for a few minutes after a cloud impacts the solar field.
3) **The power block**: The thermal storage system seen in Figure 3.10 is a two-tank (cold and hot) molten-salt system, and Section 3.6 describes how it works. Here, a steam Rankine cycle is used to transform the thermal energy from the solar field or the storage system into electricity. The maintenance requirements are the same as or very similar for conventional power plants because the power block for a parabolic-trough system includes components such as water pumps, wet cooling systems, steam turbines, electricity generators, de-aerators, and water/steam heat exchangers. These solar power plants may use an auxiliary gas-fired oil heater, which is not seen in Figure 3.10, to keep the plant running when there is no solar radiation and no thermal energy in the storage system. Due to the SEGS plants' experience demonstrating that plant operation is more challenging if the gas-fired heater is positioned in the power block and

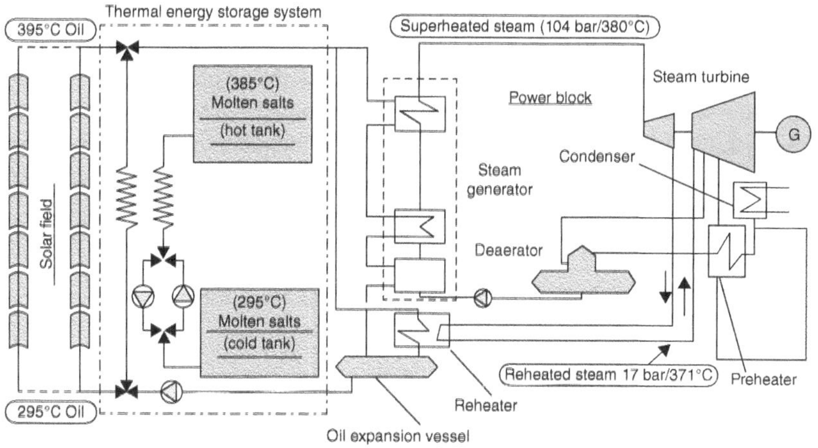

Figure 3.10 Configuration of a typical solar thermal power plant with parabolic-trough collectors.

directly generates the superheated steam for the steam turbine, this heater is often installed in parallel to the solar field. The switch from solar to fossil plant operating mode would be far more difficult, despite the fact that total plant efficiency would be higher with this arrangement since it prevents heat losses in the oil circuit.

3.3.1 Parabolic-Trough Concentrating Solar Power (CSP) Systems

The solar field tracking begins on a normal clear day when the direct sun energy is between 100 and 300 W/m^2. The thermal oil is transferred to either the thermal storage system or the steam generator to start the power block after being circulated through the solar field several times to bring it to the nominal outlet temperature. The solar field of plants with a thermal storage system produces enough energy during summer daylight hours to maintain the power block's full load while also charging the storage system. The thermal output of the solar field on clear winter days is much lower than in summer because the solar collectors are typically installed with their rotation axis facing north-south, and thermal storage is not frequently used because the thermal output of the solar field on clear winter days is only sufficient to feed the power block. Due to this, only partly cloudy days in the winter are often used with the thermal storage system. The success of the SEGS plants established it as the most financially feasible, "least-risk" technology, and the profitability attained in a few nations as a result of public incentives in the form of feed-in tariffs or tax credits implemented during the first decade of the twenty-first century encouraged a multitude of solar thermal power plant projects, the majority of which used parabolic-trough collectors. While many of the projects that were first proposed in 2008 never materialized due to the global financial crisis, many of them did, and by the end of 2010, a sizable number of CSP facilities with parabolic troughs were operational. The list of CSP plants with parabolic troughs that were in operation and producing a combined 876 MW_e of electricity is shown in Table 3.5. Such information could not be included in Table 3.5 because of the fierce industrial competition, which makes the majority of the owners of these plants rather unwilling to submit annual performance data.

3.3.2 Design of Parabolic-Trough Concentrating Solar Power (CSP) Systems

The most crucial PTC (parabolic trough collector) settings are explained in this section, along with suggestions for improving performance. The PTC energy balancing and solar field design standards are also covered.

3.3.2.1 Basic PTC Parameters

Commercial PTC designs for solar thermal power plants range in length from 100 to 150 m, have a parabola width of around 6 m, and offer an aperture area of 550–825 m^2 (approximately). Larger PTC designs are being developed and could be made available commercially in the near future.

Parabolic-trough collectors need solar tracking systems in order to adjust their location in accordance with the shifting apparent sun position in the sky from sunrise to dusk. This kind of solar collector can only move in one direction, on-axis rotation. If the rotation angle is off, the concentrator will not be able to properly reflect and concentrate the solar radiation beam onto the receiving tube. In order for direct solar radiation to correctly reflect onto the receiver tube, it must first reach the collection aperture plane (Figure 3.11). The sun vector, the collector focal line, and the vector perpendicular to the collector aperture plane must all be on the same plane for the PTC to be in the proper position. The incidence angle is the angle formed by the two vectors in Figure 3.12. It has a significant impact on how much solar flux is incident on the PTC aperture plane because more solar flux is reflected and transformed into useable thermal energy in the receiver pipe at reduced incidence angles. This portion of the solar radiation is worthless to a PTC because the concentrator cannot reflect it onto the receiver tube since diffuse solar radiation falling on the Earth's surface has no discernible direction.

The geometric concentration ratio, acceptance angle, rim angle, and peak optical efficiency are the key PTC parameters. The next paragraphs provide explanations of these parameters.

The ratio between the collector aperture area and the total absorber tube area is known as the geometric concentration ratio, or C_g. Although the maximum concentration ratio is on paper, it is typically about 25.

Table 3.5 List of CSP plants with parabolic troughs in operation at the end of 2010.

Plant name	Nominal power (MW_e)	Location	Owner	Remarks
Solar Energy Generating Systems (SEGS-I, IX)	354	California, USA	Sunray Energy, (SEGS I + II)	Collection of nine units ranging from 14 to 80 MW_e and operating since the 1990s
			FPL Energy (SEGS III–IX)	
Solnova-I, III, and IV	150 (3 × 50)	Seville, Spain	Abengoa	Solnova I (May 2010)
				Solnova III (May 2010)
				Solnova IV (August 2010)
Andasol I and II	100 (2 × 50)	Granada, Spain	ACS	Andasol I (2008)
				Andasol II (2009)
Nevada Solar One	64	Nevada, USA	Acciona Solar Power	Completed in June 2007
Ibersol Puertollano	50	Puertollano, Spain	Iberdrola	Completed in May 2009
Alvarado I	50	Badajoz, Spain	Acciona Solar Power	Completed July 2009
Extresol I	50	Badajoz, Spain	ACS	Completed in February 2010
La Florida	50	Badajoz, Spain	SAMCA	Completed in July 2010

(Continued)

Table 3.5 (Continued)

Plant name	Nominal power (MW_e)	Location	Owner	Remarks
Archimede Solar Power Plant	5	Siracusa, Sicily, Italy	ENEA	ISCC with molten-salt heat storage Completed in July 2010
Saguaro Solar Power Station	1	Arizona, USA	Arizona Public Services	2006
Keahole Solar Power Plant	2	Hawaii, USA	Keahole Solar Power	Inaugurated in December 2009
La Risca	50	Badajoz, Spain	ACCIONA	Completed in July 2009
Planta Termoel, de Majadas	50	Caceres, Spain	ACCIONA	Completed in November 2010
La Dehesa	50	Badajoz, Spain	SAMCA	Completed in July 2010
Hassi'R Mel-1	20	Laghoutat, Algeria	ABENER + NEAL	150 MW_e ISCCS plant with 20 MW_e solar. Completed in Summer 2010
Kuraymat	40	Kuraymat, Egypt	NREA	470 MW_e ISCCS plant with 40 MW_e solar. Completed in December 2010

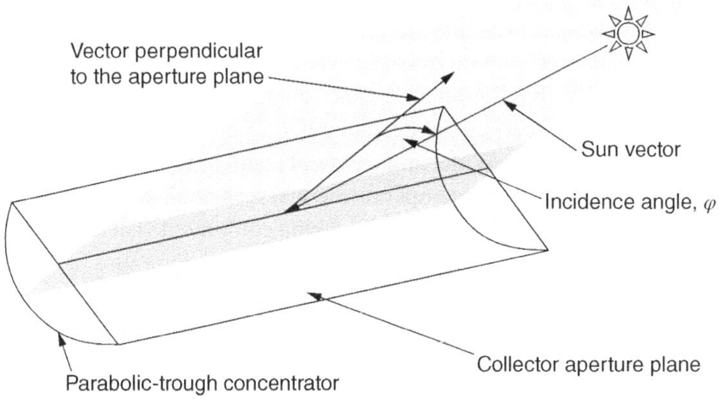

Figure 3.11 Correct positioning of a parabolic-trough concentrator.

Higher operating temperatures are related to high concentration ratios. The following equation yields the geometric concentration ratio, C_g:

$$C_g = \frac{W.l}{\pi.d_o.l} = \frac{W}{\pi d_o}$$

where W is the parabola width, l is the collector length, and d_o is the receiver steel pipe outer diameter. The acceptance angle Θ, is the greatest angle that two rays on a plane perpendicular to the collector aperture can generate in order for the reflected rays to intercept the absorber pipe when they reach the parabolic mirrors. The sun tracking system needs to be less precise the broader the collector

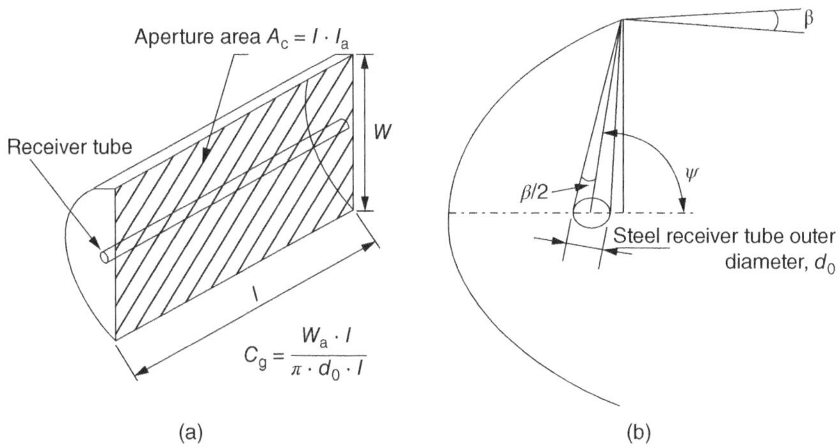

Figure 3.12 (a) Geometric concentration ratio, C_g and (b) acceptance angle, and aperture angle, of a parabolic-trough collector.

acceptance angle is, as the collector won't need to update its location as frequently. High concentration ratios are linked to small acceptance angles, necessitating the installation of very precise solar-tracking devices at a higher cost. The average solid angle from which the solar disk is seen from Earth is 32 (0.53°), which is the lowest acceptance angle. Any PTC that has an acceptance angle less than 32 would thus constantly lose some of the direct sun light. In actuality, acceptance angles of 1°–2° are advised for commercial PTCs. Wider angles would result in smaller concentration ratios and, hence, lower working temperatures, whereas smaller angles would necessitate highly precise and expensive solar-tracking equipment. Therefore, the majority of commercial PTC designs have geometric concentration ratios of 20–30 and acceptance angles of 1°–2°.

Optical losses in a parabolic-trough collector: From the following equation, the rim angle, which is a function of the parabola focal distance, f, and aperture width, W, may be computed as a function of the concentrator arc length:

$$\frac{3.f.W}{W^2 - 16.f^2} = \tan\Psi$$

Rim angles typically range from 70° to 110°. Because they reduce the aperture surface, smaller rim angles are not recommended. Rim angles greater than 110° are not economically viable since they need for a larger overall reflecting surface without appreciably expanding the aperture plane's surface area. Because optical losses account for roughly 25% of the total solar radiation incident on the PTC aperture plane, they are particularly significant in parabolic-trough collectors. The following four variables are connected to optical losses.

- **The collector reflective surface's reflectivity**: Only a portion of the incident solar flux is reflected toward the receiving tube because the PTC's reflectivity is less than 1. Mirrors made of clean, silvered glass often have reflectivity values of 0.93 or below.
- **The intercept factor**: Inaccuracies in the PTC shape, mechanical deformation of the PTC, flexible bellows, or microscopic imperfections of the reflectors prevent some of the direct solar radiation from being reflected by the mirrors from reaching the active surface of the receiver pipe.

If microscopic flaws are ignored, three factors—geometrical errors in the PTC shape, shadowing by the flexible bellows, and mechanical deformation (i.e. bending and torsion) of the support structure—can be thought of as the cause of the intercept factor with a 0° incidence angle of the beam solar radiation. All of the aforementioned factors either cause some rays to reflect at an incorrect angle or block part of the reflected rays, preventing them from reaching the steel absorber tube. The intercept factor, globally quantifies each of these losses. Because 0.97, 0.96, and 0.99, this optical parameter often falls between the 0.91 and 0.93 range for high-quality PTCs.

- **The glass cover's transmissivity**: To minimize thermal losses, the steel receiver tube is put into a glass cover. Only a small portion of the solar energy that is directly reflected by the mirrors onto the receiver pipe's glass cover can pass through. The transmissivity of the glass is defined as the proportion of radiation that passes through the glass cover to all radiation that incident on it. It is normally 0.93, although antireflective coatings used on both sides of the glass cover can raise it to 0.96.
- **The receiver selective coating's absorptivity**: This parameter measures the energy that the steel receiver pipe has absorbed relative to the total radiation that has reached its outer wall. For receiver pipes coated with a cermet selective coating, this parameter is normally 0.95; for pipes coated with black nickel or chrome, it is slightly lower.

When the incidence angle of the solar radiation onto the PTC aperture plane is 0°, multiplying these four parameters—reflectivity, intercept factor, glass transmissivity, and absorptivity of the steel pipe—gives the peak optical efficiency of the PTC, opt,0°:

$\eta_{opt,0^0} = \rho \times \gamma\tau \times \alpha|_{\varphi=0^0}$high-quality parabolic trough collectors, opt,0° is typically in the range of 0.74–0.73.

It accurately depicts the proportion of the beam's solar radiation that, at 0° incidence angle, is finally absorbed by the receiver pipe when it takes into account the four optical factors that make up the peak optical efficiency.

The four optical parameters described above as well as the useable aperture area of the collector are all impacted by the incidence angle, or, of the solar radiation beam. The incidence angle modifier, $K(\Theta)$, which accounts for all optical and geometric losses in a PTC caused by an incidence angle greater than 0°, serves to quantify this effect. The peak optical efficiency, opt,0°, is multiplied by the incidence angle modifier, $K()$, to determine the proportion of the beam solar energy that is ultimately absorbed by the receiver pipe when it arrives at the PTC aperture plane with the incidence angle.

$$\eta_{opt.\varphi\neq 0^0} = \eta_{opt.0^0} K(\varphi)$$

In most cases, a polynomial equation is used to calculate the incidence angle modifier, which depends directly on the incidence angle and has values of 0 for high incidence angles (85°) and 1 for incidence angles below 0°. So, for an LS-3 PTC, for example, the incidence angle modifier is provided by

$$K(\varphi) = 1 - 2.23073E - 4 \times \varphi - 1.1E - 4 \times \varphi^2 + 3.18596E - 6 \times \varphi^3 - 4.85509E - 8 \times \varphi^4 \; (0^0 < \varphi < 80^0)$$

$$K(\varphi) = 0 \; (0^0 < \varphi < 80^0)$$

In parabolic-trough collectors, thermal losses are also crucial since they have a significant impact on the total collector efficiency. Thermal losses overall in a PTC.

Radiative heat losses from the steel receiver tube, Qabsorber_ambient, as well as convective and conductive heat losses from it to its glass cover, Qabsorber_glass, are what cause Qcollector_ambient. Despite being controlled by the well-known mechanisms of radiation, conduction, and convection, these heat losses are experimentally estimated for each PTC and receiver pipe design by running the collector under actual sunlight circumstances at various temperatures. The thermal loss equation as a function of the temperature of the steel receiver tube and the surrounding air is then calculated using these experimental results. The working fluid temperature, the surrounding air temperature, and the sun flux incident onto the aperture plane are sometimes used to compute thermal losses. In any event, the outcome is a mathematical equation for computing overall thermal losses in the PTC, Qcollector_ambient, that the PTC manufacturer provides to the solar field designer.

The glass cover of modern high-temperature PTCs is evacuated, preventing convection losses between the steel pipe and it.

3.3.2.2 Energy Balance in a PTC

A typical PTC's energy balance is shown in Figure 3.13. On the left is the solar energy flux that strikes the PTC's aperture plane (Qsun_collector), and on the right is the thermal energy that the PTC delivers that is useful (Qcollector_fluid). The three causes of energy loss in the PTC are represented at the bottom of Figure 3.13 and are described in Section 3.4.1 as follows:

- Optical losses resulting from the receiver tube's mirror reflectivity, intercept factor, glass transmissivity, and solar absorptance at 0° solar radiation incidence angle, opt,0°
- Additional optical and geometrical losses caused by an incidence angle larger than 0°, known as *K*(); these extra losses are absent when the incident angle is equal to 0° since *K*(0°) 1.

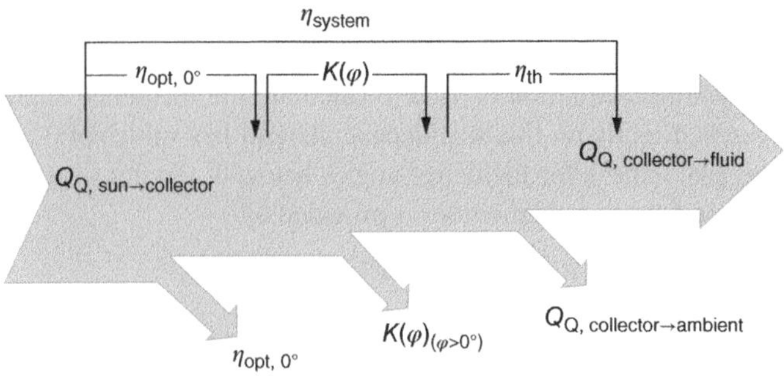

Figure 3.13 Energy balance in a parabolic-trough collector. Design and optimization of central receiver systems.

- Thermal losses in the ambient air from the receiver pipe, measured by Qcollector_ambient.

The considerable capital cost of the extremely big capacity systems, which many commercial proponents believe to be more cost-effective than small or modular systems, has to some extent slowed down the commercialization of Central Receiver (CR) systems. In recent years, smaller systems have been designed and built in order to take advantage of a lower capital cost requirement, smaller and therefore easier to handle components, quicker learning/experience curve progression, easier access to capital, quicker construction time, and potential use of readily available off-the-shelf components. It will be interesting to see if these advantages can overcome the economies of scale offered by larger systems. A recent unmanned, 1 MW_e trough-thermal-electric system that operated successfully offered hope for future unmanned, tiny CSP systems. Unmanned operation is a crucial feature for making smaller systems cost-effective. To accommodate larger applications and to save operating and maintenance (O&M) costs, such tiny systems can be clustered. However, it is important to carefully analyze the costs of energy transfer to the central site because, after taking these costs into account, using fewer larger units might be more cost-effective.

3.3.2.3 The Objective Function for Optimization

A CR plant's designer is required to keep some design parameters in mind. Typically, a system trade-off or optimization procedure will follow. Because collectors are expensive, one may choose to maximize plant efficiency, which is achieved by raising the output temperature to improve Carnot efficiency. However, doing so will probably necessitate the use of more precise (and potentially more expensive) heliostats, as well as higher thermal losses, optical spillage, and receiver material costs, as well as a generally more expensive power block, even though higher efficiency can lower the cost of auxiliary expenses like cooling. However, some cycles, like the supercritical CO_2 Brayton turbine, offer extremely high efficiency at temperatures that are quite attainable. On the other hand, choosing the least expensive option may lead to the selection of less precise heliostats, a larger receiver to compensate for spillage, higher thermal losses or a lower operating temperature, and poor system performance, possibly leading to a higher levelized electricity cost. Another option is to construct the plant for a specific day or time, which would lead to relatively poor performance at other times and over the course of the year. We assume that a power plant or a process heat application is the consumer of the thermal energy generated by a commercial CR plant. The precise required power and capacity factor, as well as a predicted temperature range, will then be determined by the application. The solar farm is then created to comply with at the lowest cost per unit of energy for this annual demand. The appropriate objective function is a cost–benefit ratio, which is the lifetime cost per MWhr of the thermal energy delivered, divided by the net output of the plant over its design lifetime. This includes the total capital cost of the installed plant as well as the present value of operations and maintenance over the projected life of the plant. It is significant to highlight that the optimization process should make use of multiyear average diurnal insolation. A single chosen year is unlikely to be representative, and while a “typical year” can be created, it is simpler to create and comprehend an insolation model based on long-term atmospheric properties. Levelized cost of electricity (LCOE), which is described in more detail in Chapter 2, is an alternative objective function that is widely employed. LCOE is defined as the unit cost at which the product must be sold in order to produce a certain rate of return throughout the life of the plant. Because the financial aspects must be defined thoroughly and into the future and have a significant impact on the outcome, this analysis is more complicated. Additionally, the cost and performance information for the plant’s nonsolar balance is needed. Few if any of these problems are solar-related, and while LCOE is a number that utilities and banks may easily recognize, financial numbers change significantly over time, and it may be challenging to get and implement meaningful performance details for the subsystems. A financial analysis method that can model the advantages of time of day dispatch (where a kWh of electricity may be worth more in the afternoon/evening than in the middle of the day, for example) will, however, be of most value to investors given the emergence of dispatchability as a key benefit of CSP.

To accept the diurnal efficiency curves for the optimum fields generated by programs like RCELL or DELSOL, SANDIA has created codes like Solergy, which do an in-depth estimate of the energy or electricity produced during a typical or representative year. The thermodynamic quantities and financial criteria necessary to create an LCOE have been taken into account by Solergy.

It is necessary to take inflation’s effects into account. Since different components or subsystems may be priced at different times, it makes sense to multiply each price by the appropriate inflation rate. The well-known consumer price index provides a general method for accommodating inflation, but at the component level, an engineering source like the Chemical Engineering Plant Cost Index (CEPCI, published monthly by the American Society of Chemical Engineers) may be a more appropriate method. If each cost is multiplied by the current index value and divided by the index value appropriate for the date it was generated, costs generated at different times can be compared and/or used coherently in a study. ‘Legacy’ is made possible. Costs from previous work will be combined with current costs. Such resources are present in many nations. The incorporation of foreign exchange rates may become a more complicated factor. Items to include in the cost function: Money invested during the course of the plant's life becomes the numerator in the objective function. Preliminaries and fixed, acquisition and construction, and finally operations and maintenance (O&M) can all be categorized under this.

Fixed costs such as permitting, design, and access: These should cover the majority of general expenses like plant design, environmental and political permits, central facilities like control rooms, fire and safety facilities, and maintenance shops, as well as access roads and transmission lines to the site, which could be expensive if the site is remote. Each of these amounts can cost more than a million dollars; thus, they need to be carefully studied and distributed among the facility's numerous subsystems. In reality, they are a major factor in the higher cost of energy produced by tiny, standalone facilities. If several small plants are integrated or co-located with one another or the energy customer, this problem might be somewhat reduced.

Capital costs: The heliostats, receiver, tower, thermal transport system, and thermal storage system, if included, are the key solar-related capital costs of a CR plant. Each of these must also include the installation costs (shipping, foundations/support structures, building or installation, connection to power and control system, quality control, and testing), in addition to the item's purchase price. A heliostat may cost less than half as much in the factory than it will when it is installed, wired, aligned, tested, calibrated, and in use at a different location. Similar to horizontal pipe, the cost of vertical piping must include for delivery to the location, hangers, a budget for expansion bends, on-site welding and inspection, insulation, heat tracing if necessary, temperature sensors, valves, etc. The real cost of the manufactured active receiver (tubes, headers, etc.) represents only a small portion of the receiver's overall cost, which is a common surprise but significant fact. The sturdy backs and hangers that the panels are attached to, the support structure, insulation, temperature sensors, valves, and controls, header shields, and ovens (if required), as well as the transition from the tower to the support structure, account for the majority of the cost. These expenses even greater for a cavity receiver that also needs to keep the reception tubes and headers inside of a high-temperature container. The cost of each subsystem must be specified up front in terms of the necessary design variables in simple relations that are true over the range of realistic values in order to enable optimization of the CR plant. Towers of a few heights, for instance, can be created using designs adequate for the local earthquake regime and top loading depending on anticipated receiver size, weight, and wind speeds. After any obvious fixed or linear costs (such as soil testing, safety lights, and elevators) are taken into consideration, the tower cost for any height within a suitable range can then be approximated with a power law or exponential fit. The tower design and cost will, of course, be more accurate and detailed in the final design.

Land: A good solar site will probably be relatively remote, with peak direct-beam insolation of around 1 kW/m^2 and annual beam insolation (DNI—direct normal insolation) of preferably around 2,000 kWh/m^2/year. It will also likely be of an appropriate size (including a buffer), moderately level (with less of a need for central receivers than troughs), and suitable for construction. A willing seller, good and assured access to adjacent transmission lines that can absorb the power, suitable roads to handle construction equipment, and readily handled environmental constraints are also required. Such perfect land might sell for \$1–2/m^2 (about \$4,000–8,000 per acre) in the United States. Even while much less expensive land might be available, meeting or dealing with all the aforementioned requirements might be exceedingly costly, time-consuming, and even impossible. Although accommodating, gently rolling soil will make heliostat layout quite difficult. Sloping land, up to perhaps 10°, can be accommodated in the system design of a central receiver plant with various effects, including: later sunrises/earlier sunsets due to virtual horizons, shifting the centroid of the field downhill, or moving the diurnal energy output curve toward morning hours if the slope is up to the west or to better meet an afternoon peak load if the slope is up to the east.

Heliostats: Heliostats must be rigid enough to withstand gravity and wind loads while retaining precision (beam error less than 2–3 mrad) in order to prevent excessive degradation of the solar picture. These heliostats are currently projected to cost around \$200/m^2 in 2010 US dollars, and if volume production and learning curve effects start to take effect, most cost targets are between \$100 and \$130/m^2. The DOE SunShot long-term goal of \$75/m^2 will be challenging to meet.

The ideal heliostat area is difficult to pinpoint because it necessitates a trade-off between various factors, including spillage (the size of the beam at the receiver), the number of control systems that must be installed and maintained, and reflector support deflections under wind and gravity loads. 'Commercial' heliostats currently range in size from 1 to 150 m^2; however, designs up to 200 m^2 have been developed, and some businesses have thought about ganged or autonomous heliostats that are much smaller than 1 m^2. A heliostat needs to improve the plant's financial performance on a net positive basis in order to be cost-effective. For instance, if the land cost associated with a heliostat is 2% of the cost of the heliostat near the center of the plant, the effective land cost will be 6% near the field boundary where the density of the heliostats is typically one-third that at the field center. Similar to spillage losses, atmospheric attenuation occurs between the heliostats and the receiver. The wider radial separation away from the receiver has a detrimental effect on the cost of conventional wiring and control systems as well. Thus, to optimize the objective function, trade-offs must be made between the location of the outer boundary and heliostat losses and related expenses.

3.4 Operations and Maintenance (O&M) Costs

The yearly charges for capital costs may equal one-fourth of the annual costs for O&M. Therefore, it is crucial to factor these expenses into the optimization. The O&M cost, which consists of both fixed and variable components and varies greatly from subsystem to subsystem, makes it improper to simply multiply the overall plant cost by 125%. For instance, the Utility Study from the 1980s used the following consensus estimates for the yearly O&M in relation to each system's construction cost: tower 0.2%; receiver 2%; vertical pipe 1%; feed pump 5%; and heliostats 1%. The reflectors' area and the tracking system's number were equally expensive for the 39 m^2 heliostat that was under consideration at the time. The argument remains the same, but the experience gained since then will enable the development of new operational values for each subsystem. Be aware that in order to compare these percentages to capital expenses, a present value factor must be used. Depending on the real rate of return assumed, this factor can range from 10 to 20 times the yearly costs.

3.4.1 Choice of Performance Criterion

Design point or annual costs are in the numerator of the objective function for the best design of a CSP plant and performance is in the denominator. There are several options for defining the performance, some of which are significantly more effective than others. Choosing performance at the design point will lead to a lower-cost system with excellent performance at that time, but relatively poor performance at other times throughout the year, and consequently a lower annual energy output. While this may be suitable for a solar furnace with short seasonal operating times, it is inappropriate for a power plant. Here, the requirement is the lowest LCOE, so the denominator must be delivered annual energy.

3.4.2 Incident, Absorbed, or Delivered Energy

The energy incident at the receiver is diminished by the receiver's interception and reflectivity (absorptivity), radiation and convection losses (although conduction losses are negligible), piping thermal losses, and parasitic losses of the feed pumps, valve, and heliostat actuators, etc., necessary for the operation of the solar complex (converted from electrical to thermal by multiplying by the heat rate of the turbine or charged directly as purchased electricity). The net-delivered thermal energy that can be used by the turbine or to charge a thermal storage unit is the outcome.

3.4.3 Inclusion/Effect of Time-of-Day Pricing, Sloped Fields

If time-of-day pricing is in place, the plant designers can exploit it to increase the plant's annual revenue by setting up the field layout to be more efficient during the period of peak electricity price. You can use this whether or not there is storage. Prior to optimizing the field, it is more effective to value the sunshine proportionally to the price associated with the time of day. As a result, high-value sunshine will be more important in determining heliostat spacings worldwide and the plant's perimeter. This will result in an east-biased field, in terms of both heliostat density and field boundaries, with an afternoon peak price. In the event of an afternoon peak price, a field that slopes up to the east will have a similar impact and is preferred. A little more efficient field will result from an upslope to the north (or south in the southern hemisphere); however, the cost of heliostat installation on a sloped field must be taken into consideration, as well as a higher polar bias of the field.

3.5 Effect of Constraints on Optimization

Any design restriction will have a detrimental effect on the field or system that results from optimization, and this impact will be even more pronounced if the restriction is put in place after the optimization. The lowest LCOE is the goal of the unconstrained optimization process. As a result, it specifies the heliostat spacing and number (field boundary) for a particular tower height and receiver aperture and assesses the design point power and annual output power. Trimming heliostats from the field's edge can result in a lower design point power and perhaps improved efficiency, but as cost-effective heliostats are eliminated, the LCOE will rise. The optimization procedure will result in a lower tower, a smaller receiver, and an intermediate area field, all at an intermediate LCOE, if the design point power is alternatively constrained from the outset. It goes without saying that the system with the fewest constraints will be the most economical, and the optimal moment to introduce any constraints is before the optimization. It is obvious that the definition of a polar-facing aperture is a strong constraint, but the constraint can be partially eased by using two or three chambers to increase the azimuthal acceptance angle.

3.6 Heliostat Factors

Beam errors: An image of the sun can be formed on the receiver at a distance of one focal length using an optical mirror that is perfectly focused and functioning on-axis. The picture won't be a point since the sun is a far-off object with an angular diameter of 3.306 mrad; instead, it will have a diameter of 0.0093 times the focal length. This would have a diameter of 3.3 m at 1,000 m and show the sun's limb darkening (caused by scattering in the solar atmosphere) and solar aureole (light that appears outside the sun's limb owing to scattering in the Earth's atmosphere). Instead of focusing on creating a picture for CSP systems utilizing concentrator optics, our goal is to efficiently capture the majority of the energy. As a result, it is possible to conveniently describe the sun as a Gaussian function that is a least square fit to the limb-darkened solar picture, giving rise to a second moment (sigma) of 2.770 mrad but overestimating the peak flux density by roughly 16%. The resultant Gaussian 'spot' at 1,000 m will have a diameter of 5.54 m at the one-sigma radius (relative intensity = 0.607) and a diameter of 3.08 m at the two-sigma radius (relative intensity = 0.135). If the Gaussian is a good approximation to the solar aureole, the energy beyond this diameter (i.e. the spillage for a circular target) would be mathematically equal to the relative intensity at the defined radius. If beam errors are comparable to the second moment of the sunshape, the improvement is marginal and the complexity is significant. A better fit to the true sunshape can be achieved by adding to the second moment (sigma2) higher moment terms such as fourth moment (a measure of kurtosis) and sixth moment in a Hermitian series. It is helpful to include words up to sixth order in the Hermite series for extensive flux analysis. All heliostat system flaws are referred to as beam errors, and they are quantified in terms of the beam's half-angle of divergence when lit by a real point source. Since it is believed that all systemic problems are detected and fixed by standard O&M methods, the remaining errors, at least when averaged over a collection of heliostats, are essentially random. Reflection magnifies any remaining mechanical flaws that could change the mirror's orientation. These mistakes include tracking mistakes, wind or gravity impacts on the heliostat, etc. It is incredibly convenient to manage all of these random faults by adding them in quadrature to the second instant of the limb-darkened sun (2.770 mrad) to create a "degraded sun" (after multiplying mechanical errors by four). By doing this, all the problems associated with converging several errors on numerous heliostats are avoided. This also offers a helpful benchmark for determining the significance of beam errors: if the total beam error is equal to that of the sun (2.770 mrad), it will result in a degraded sun with a sigma of 1.41*2.770 3.92 mrad; otherwise, it will be dominant. If the beam error is much less than this, it will be relatively unimportant. Since we typically have thousands of heliostats, we can regard individual instantaneous displacements of specific beams or facets as random variables, lending credence to the idea of a damaged sun. However, it does sacrifice the capacity to forecast accurate images from a single heliostat. The aforementioned Hermite approach also allows for the incorporation of higher-order moments of the sun and of the mistakes.

3.6.1 Heliostat Size

The optical justification for taking into account heliostat size is provided by the discussion of sun-shape and beam errors. Simply said, an optically flat heliostat will project an image of the sun from each point, creating a mirror image with umbra and penumbra effects that correspond to the solar beam. The intensity of the picture will not be greatly diminished compared to a focused mirror as long as the mirror's angular size as seen from the receiver is less than that of the sun. There will be a significant reduction in intensity from a flat mirror compared to a focused mirror if the mirror is comparable to or greater than the projected solar image, along with a corresponding increase in the spot size.

3.6.2 Focusing and Facet Canting

By splitting the mirror into facets (which also makes them easier to handle) and canting each one to superimpose their pictures at the receiver (which comes with some cost in hardware and labor), the spot degradation produced by large, flat mirrors can be completely recovered. Alternately, to lessen the impact of the facet diameter on the sun spot, the mirror or individual facets can be bent to focus their beam close to the receiver surface (often referred to as "to slant range"). Due to Poisson ratio effects, it is significantly simpler to bend glass into a cylinder than a sphere. Regardless of the optical geometry , the image of the error-free mirror can be projected to the receiver via ray tracing, and the second moment (and higher moments) of this "spot" can be calculated and added in quadrature to that of the degraded sun, or the image can be generated by detailed ray tracing.

3.6.3 Off-Axis Aberration

Off-axis aberration is a phenomenon that can be caused by either canting or focusing. Since the angle of incidence () of the sun's rays on the mirror effectively modifies the focal point, the focal length (f) is reduced in the transverse direction by cos () and raised in the sagittal direction by 1/cos (), generating two line foci with a circle of least confusion between them close to f. The diameter of this circle is equal to $w(1 \cos i)$ for a mirror with diameter w and incidence angle i. This results in a heliostat image from a parabolic (or small spherical) mirror that is "perfectly focused on axis to slant range" at an off-axis angle of 60°, which is added in quadrature to the sun image and the effects of beam faults explained above. As one moves further away from the nominal focal length, the situation becomes more challenging.

3.6.4 Effects of Tracking Mode

The tracking of the heliostat image onto the receiver can be done using various mounting solutions. The most popular system is the elevation-azimuth (el-az) one, which rotates the assembly in azimuth while mounting the orthogonal elevation axis carrying the mirror above it. Although easy to construct and inexpensive, this plan has a number of drawbacks. One is that as the reflector moves in azimuth, it loses its orientation with respect to the plane containing the sun, mirror, and receiver. As a result, it is generally agreed that the optical figure of a sphere for the (almost flat) mirror is the optimum choice because the angle-of-incidence effects discussed above cannot be resolved. To maintain the proper orientation with regard to the sun-mirror-receiver plane, the mirror can also be provided a third axis of rotation (centered on its center) and constantly rotated. This can also be done using a spinning-elevation tracking system or a tracker with the first axis oriented in the direction of the receiver. The cos effect can then be eliminated by applying a suitable off-axis parabolic optical figure to the mirror. In any case, el-az mirrors need a "clear-out circle" with a diameter equal to the maximum dimension of the horizontal mirror (D for a circle, 1.41 S for a square) to prevent collisions with its neighbors in the event that tracking control fails for a rectangle, (L2 W2)0.5, etc. As a safety measure, typically 30 cm (for a large heliostat) or more is added, depending on whether the axis of rotation passes through the center of the horizontal mirror or not. This "mechanical limit" could make heliostat separation requirements more stringent, which would be problematic in the most productive area of the field where heliostat density is typically highest. Theoretically, a computer algorithm could prevent collisions between heliostats that are positioned closer together, but failures would be expensive. A mirror that does not revolve in azimuth is mounted on an elevation/polar axis mounting mechanism. When compared to the el-az mount, this enables mirrors to be packed together, resulting in twice the maximum mirror density for square mirrors. Additionally, it moves continuously around the polar axis. For modest adjustments to account for seasonal variations in solar declination, use the orthogonal axis. A mounting mechanism can rotate at a constant speed with the first axis facing in the direction of the receiver so that the second axis is constantly perpendicular to the direction of the sun. In this scenario, the mirror will always maintain the proper orientation in relation to the plane formed by the sun, mirror, and receiver. There won't be a cos impact if the optical figure of an acceptable off-axis parabola is provided. Before choosing these and other unique mounting techniques, it's crucial to weigh the extra expense against the benefits.

3.6.5 Effects of Heliostat Size on Heliostat Cost and Other Factors

When choosing the size of the heliostat, all of these impacts must be taken into consideration. Clearly, the concentration and spillage are impacted by heliostat dimensions. Additionally, larger heliostats (to a degree) benefit from economies of scale, but smaller heliostats (which will need more pedestals, controls, actuators, etc.) gain more from learning/experience curve effects. There is currently no agreement on heliostats from 1 to 200 m^2. Smaller heliostats will undoubtedly help tiny systems, but O&M expenses must also be taken into account.

3.6.6 Reflectivity and Cleanliness

Of course, the imperfect reflectance of all heliostat mirrors results in losses. This can range from 3% for a front surface silvered mirror with good protection to 4% for a second surface mirror made of water-white glass to 15% for thicker, high-iron glass. This perfect reflectivity will be very constant over time for any suitable mirror, but by the time it is put in the field, it may have suffered a 1% or so degradation due to surface dents, scratches, etc. The cost/m^2 of improved reflectivity should be compared to the cost/m^2 of the heliostat field (usually five times bigger), not to the cost/m^2 of the mirror, as any fall in reflectivity reduces the heliostat's performance.

Once mounted, corrosion-free mirrors have a 2% monthly reflectivity loss that is caused by a combination of surface layer that is more difficult to remove and easily removable dust, depending on the location and climatic conditions. Since cleaning equipment is expensive and labor-intensive, it is necessary to balance the cost of washing frequency against the cost of reflectivity loss because zero loss means infinite cost. While scrubbing to remove the surface coating requires a slower and more expensive brushing gear, rinsing to remove the dust may be done with a boom truck and is relatively quick. In contrast, if a less expensive rinse cycle is added to simply remove the dust between scrubbings, the loss in average reflectivity with optimal cleaning may fall to 3.5% (depending on the environment, rig costs, and labor costs). If the mirrors are scrubbed at the recommended rate, the loss in average reflectivity may be limited to about 6%. An ideal average reflective loss can be found by balancing reduced washing expenses against the price of the more heliostats needed for the filthy field to provide the same energy. Production objectives won't be attained until the optimal average reflectivity is used to calculate the necessary size of the heliostat field.

3.7 Receiver Considerations: Cavity vs Flat vs Cylindrical Receivers

3.7.1 Field Constraint

A common design restriction is to specify a restricted receiver aperture, such as a hollow or billboard receiver instead of a cylindrical receiver. A billboard's field of view is 180°, at most, compared to a cylinder's 360°. However, by tilting the aperture forward, more ground can be seen. Near noon, when the majority of the heliostats experience near 0° angle of incidence of the sunlight and have a "cosine effect" near unity (a combination of the foreshortening of the heliostat by cos and off-axis aberrations), a polar-facing billboard with a 120° field of heliostats will perform very well. The performance declines quickly, though, as the sun azimuth changes (during other hours of the day). Contrarily, a system using a cylindrical receiver will perform moderately at midday (as a result of the decreased cosine that heliostats on the equatorial side of the receiver experience). However, the nearly circular shape of the surrounding field tends to maintain the average value of the cosine over the field as the azimuth of the sun moves away from its noontime value (by more than 100° in the summer), so the annual performance for the surrounding field outperforms that for the polar field of the billboard or cavity. A cavity may be necessary to reduce thermal losses for high-temperature working fluids with low heat transfer coefficients (gases), but for small systems, a billboard may be much less expensive than a cylindrical receiver. It is important to consider how the resulting constraint may affect field performance.

3.7.2 Reflective, Radiative, and Thermal Loss of the Cavity

It is untrue that a cavity's entrance is a "black hole." In reality, one benefit of cavities is that excessive flux density problems can be minimized by distributing the sunlight that enters the aperture over the absorbing surface. A large portion of the sunlight is typically reflected or dispersed off adiabatic surfaces to achieve this process. Due to this, a lot of diffuse solar light bounces around in the cavity, some of which hits the aperture and is successfully reflected back out into space. The fact that a large percentage of the cavity's interior absorbs and reradiates light at a temperature higher than the heat transfer surface and that a sizeable amount of this infrared energy is likewise emitted through the aperture is a secondary effect.

Finally, the entire cavity interior comes into contact with the contained air, raising its temperature to an extremely high level and causing turbulent convective circulation rolls. Such circulation, which is heavily promoted by ambient wind, can convey heat to the aperture and out depending on how the cavity is oriented. As a result, a hollow receiver with inadequate design could sustain larger losses than a billboard- or cylindrical-shaped external receiver. A window can be useful for small receivers, but it is challenging to scale up for commercial-scale systems.

3.7.3 Cost and Weight

For very high-temperature systems, where it is advantageous to have a small ratio of aperture to absorber area and where high fluxes can be absorbed, cavity receivers have a specific application. The support structure needed for an external receiver, where the total absorbed flux is the same, will undoubtedly be smaller than the cavity that must support and enclose the heat-collecting surface because of this. To minimize conductive losses due to its substantial surface area, it must also be adequately insulated. As a result, the enclosure for the actual receiver is bigger, heavier, and more expensive. Because the constrained cone angle necessitates a higher tower, cavity receivers are often created for systems with smaller capacities. Even for an external receiver, it is frequently discovered that the receiver tubing, headers, valves, etc. only make up a small portion of the overall assembly cost.

3.7.4 Effect of Allowable Flux Density on Design

A heliostat field with a peak concentration of several thousand suns (several MW/m^2) can be easily designed. Nevertheless, the majority of commercial system receiver designs can only withstand a flux density of 1.0 MW/m^2 (for molten salt, for example) as a result of the fluid's properties for heat transport. Therefore, a receiver that is larger than the lowest attainable focus point must be employed. To enable close-in heliostats, which view a severely foreshortened receiver, to function, the aspect ratio of an external receiver (height/diameter) should be greater than one. A large tube length is typically required by the receiver designer in order to reach design temperature at the high heat-transfer-media flow rates necessary to increase the allowed flux density. It is typical for the final receiver aspect ratio to fall between one and two. Then, by defining many (2–5) goal levels, to each of which particular fractions of the heliostats are assigned, the permitted flow is accomplished. Very little spillage will be introduced while the peak flux is reduced by 1.5–4.0 times using a clever method that assesses the beam radius and transfers the image associated with heliostats assigned to the outer aim levels so their beam only remains on the receiver.

The ratio of north to south side flux may also need to be restricted in order to preserve some balance on the receiver panels and lessen transverse gradients on each panel. This can be done by strengthening the field boundary on the "weak" side while weakening the opposite side. To 'inform' the field of the new constraint, a more advanced method entails changing the objective function of the optimizer with a function (for instance, cosine) of the azimuth angle. The density and extent of the fields are changed, maintaining the idea of an optimal field while relaxing the receiver and resulting in cost savings. At some expense, this disadvantages the polar heliostats and benefits the equatorial heliostats. A potential issue is the flux density gradient that occurs across each panel; however, this is mostly determined by the antisun to sun-side flux ratio, which may be changed as described above. Smaller images' transverse shifts can somewhat lessen the local transverse gradient.

3.7.5 Emissivity vs Absorptivity vs Temperature

Reduced emissivity absorbers (solar selective surfaces) can be employed effectively to lower radiative losses at the lower temperatures and concentrations typical in linear systems. However, it is challenging to build selective absorbers that will endure when operating at temperatures over 500°C and exposed to the atmosphere. Reducing the emissivity in this region will also reduce the absorptivity because at each wavelength, energy is emitted by the sun and a hot receiver. Between 1.5 and 2 micron wavelengths, a strong transition from 1 to 1 is necessary. It is crucial to remember that the goal is to increase the net energy captured in the receiver. At a solar flux density of 1 MW/m^2, a reduction of 0.01 (or 0.02) loses 10 (or 20) kW/m^2, necessitating a reduction from 1.0 to 0.56 (or less) at a surface temperature of 527°C (800 K). Thermal losses will rise when higher temperature receivers are predicted (for example, to power a supercritical turbine), and as the surface temperature greatly surpasses the working fluid temperature. However, as the solar and IR spectrum overlap, the task will become much more difficult.

3.8 Variants on the Basic Central Receiver System

Due to substantially reduced incidence angles on the antisun side of the receiver, a billboard or cavity receiver naturally produces a polar field (often referred to as a north field in the northern hemisphere), i.e., one with heliostats on the polar side of the antipolar facing receiver. Due to the increased values of incidence angle (decreased cosine) on practically all the heliostats, such a field will perform quite well around noon but will rapidly deteriorate at greater solar azimuths. In contrast, a surrounding field with a pole/equator energy ratio between 1.5 and 2 will be best matched to a cylindrical receiver to take advantage of the better cosine effect. At midday, this system will function less consistently than the polar field, but at big sun azimuth, both the stronger east and west fields and the weaker equatorial field will significantly increase energy absorption. As a result, the surrounding field with a cylindrical receiver has significantly higher yearly energy gathered for a given design point power and is therefore more advantageous in terms of cost per unit of energy.

3.8.1 Beam-Down Systems

A secondary mirror is inserted between the heliostat field and its raised focus point in a beam-down CRS in order to guide the solar light gathered to the ground. The mirror must be a hyperbola (or an ellipse if above the primary focal point) and large enough to intercept the full cone of light from the field in order to produce a focused spot close to the ground. A Cassegranian system is one that efficiently uses this type of system in astronomy. A tower or even a dish system is frequently met with the following common response: "If you have the receiver supported on a tall tower and the engine down on the ground, why not use a Cassegranian system with a small elevated mirror and the receiver down near the ground where it is easy to connect to the engine and to service?" Why not here if it works in astronomical telescopes? Additionally, a lot of the heat transfer pipe would be saved. The reason solar collectors frequently (generally) have an aperture diameter several times the focal length is because astronomical telescopes universally have a small f number (ratio of aperture diameter to focal length). This is due to the fact that stars cannot be resolved.

The sun, in contrast, has a diameter of 3.3 mrad, and we are attempting to create a focused "image" of it. In a Cassegranian system, the high "hyperbolic" mirror creates an image of the virtual object, or the picture that would be created by all of the heliostats' light convergent at the array's focal point. The ratio of the distance from the secondary mirror to this actual image divided by the lesser distance from the virtual object to the secondary mirror magnifies the real image created from this virtual object, which may be close to the ground. To have a small secondary mirror (relative to the area of the collector field), the secondary mirror must be close to the (virtual) focal point of the array and far from the ground (actually, the receiver aperture which may be elevated somewhat). This is because the convex-downward hyperbolic Cassegranian mirror must be large enough to intercept the entire cone of rays directed toward the focal point. The linear magnification of the assembly, which would be 10 to produce a secondary mirror area 1% of the field area (3% or 4% of the collector area due to heliostat spacing), is instead determined by the ratio of the distance from the virtual focal point to the vertex of the secondary and the distance from the secondary to the receiver aperture (which may have a small elevation). The virtual object's area is magnified by 10 times, which reduces the flux concentration by 100 (from a few thousand to a few tens). This is often bad for the goal of achieving a high temperature with little loss. The final concentration is improved by moving the secondary mirror lower, but this requires an oversized mirror and three or four support towers, which is not practicable. Moving it halfway restores full concentration at the ground (i.e. the necessary flat mirror precisely replicates the virtual image). If the average ground coverage is 25%, the secondary must have an area that is about equal to that of the mirrors or be half the diameter of the field. A significant shadow will also be cast on the field. If a beam-down configuration is required, supporting the receiver above the ground level should be taken into account because doing so diminishes magnification.

3.8.2 Use of Compound Parabolic Concentrators

Compound parabolic concentrators (CPCs) placed at the receiver aperture can help restore some of the concentration that was lost due to the beam-down design. The acceptance angle of the CPC must be set at the cone half angle determined by the secondary radius/secondary distance (from the CPC aperture) in order to accept the complete beam reflected by the secondary. The concentration of the 2D (conical) CPC is also defined by this angle, as $(1/\sin\ )^2$. A 'fly eye' array of 7 or 19 CPCs may be necessary if the secondary's final picture is large, but each one should observe the entire secondary to prevent overly harsh rejection of light outside the view angle.

A noncompetitive design is almost invariably the result of the CPC array's additional cost and complexity, the secondary's cost, and the cost of sustaining it on several (albeit smaller) towers. Of a larger angular field (as seen from the tower-mounted receiver), using a CPC with a typical central receiver system is debatable. However, at the Sandia 5 MWth test site, a number of linear CPCs were stacked to improve the flux on a long, narrow panel receiver that was being tested, with each CPC viewing a different sector of the north field (NE, N, and NW). Though it was fairly pricey, this creative approach was incredibly effective. The concentrated light at the exit aperture of a CPC is diffuse light, which means it spreads out to the full 2Pi steradians. Truncating the exit aperture, which also lowers the concentration, can help to some extent.

3.8.3 Optical Beam Splitting

When an energy conversion device's performance is wavelength sensitive, it is occasionally recommended that the energy reaching the receiver be split into two or three wavelength bands using a beam splitter to maximize the usage of each beam. One reason for doing this might be to employ a portion of the wavelengths for heat conversion and the rest for concentrating photovoltaic conversion. Numerous research initiatives exist; however, there hasn't been any commercial use yet. Installing transparent solar cells in front of the receiver with all light that is not being consumed by the PV cells being transmitted to the receiver would be the simplest use. This could be employed effectively in dish systems, although being challenging to install in big systems. The entire solar beam would be effectively absorbed by a thermal receiver because it is typically a wideband absorber; nothing would be lost. When a thermal device produces power more cheaply than a concentrating photovoltaic system, the thermal device would be responsible for utilizing the residual spectrum, assuming there is still enough energy to justify the cost. In addition, the UV rays have a low initial energy and most of this energy is absorbed during the initial reflection (due to imperfections in the glass, but this is not a problem with front surface reflection), meaning that very little energy actually reaches the receiver. At the receiver, it might be able to separate the IR from the visible beam, but it is difficult to think of a better application for this IR than to add to the energy absorbed by a thermal receiver.

3.9 Field Layout and Land Use

The system will increase the density of heliostats throughout the field, place more heliostats at the field's edge, and raise the tower's height to make up for the increase in LCOE. It is interesting to note that a well-defined, optimized, unconstrained field (with a cylinder receiver), defined by the annual output/m^2 of glass, is almost circular but is offset from the tower toward the pole by 10–20%, depending on latitude. However, a field boundary determined by output/m^2 of ground ultimately ends up being extremely close to a circle centered on the tower if land use is at an absolute premium.

Field layout for optimized systems: The optimal compromise has been determined empirically to be a circular radial-stagger (RS) heliostat configuration. The radial heliostat spacing is mostly controlled by blocking from those aligned neighbors closer to the tower when the receiver is far away. This problem is solved by the RS configuration by relocating the closest radial neighbor to two circles. The footprint of the annual darkening is more or less spherical and leans toward the pole. This results in the practically universal requirement that nearest heliostats be at least two diameters apart in azimuth. Additionally, by 'peaking through' between the neighbors in the first inner circle, the distant heliostats are able to avoid blocking. Heliostats often function nearly horizontally to illuminate the receiver close to the tower. As a result, there is hardly any blocking and little shading, and they can be placed much closer together. Here, mechanical restrictions must be put in place to stop errant heliostats from colliding with one another. Round heliostats and polar mounting solutions, which prevent the mirror from ever leaving its original footprint, are advantageous in this area. The RS configuration is rendered ineffective close to the tower due to the rapidly changing azimuthal spacing imposed by the increasing circles. One possibility is to utilize a hexagonal close-packed arrangement close to the tower before switching to the RS design and seamlessly merging the interface at a radius of one or two tower heights. Another option is to simply arrange the heliostats in rings that round the tower, spacing the circles as closely as feasible to permit maintenance and adhere to mechanical constraints, as was done at Gemasolar. Blocking will become substantial at some radius, at which point the RS arrangement can be used. In a densely packed field with polar/elevation tracking, the 'peaking through' advantage does not happen as often. At some certain design point, it is possible to construct "more optically efficient" configurations for energy production; however, they may not be the most economical and frequently ignore receiver limits.

3.9.1 Ease of Access for Maintenance

To enable routine cleaning and sporadic repairs, it is crucial to retain easy access to each heliostat. While some proposed layout concepts are extremely densely packed or virtually random (each heliostat is positioned at the 'optimal position' depending on the area left by preceding heliostats), the RS design and the east-west orientation both permit this. Field designers must realistically take installation and ongoing O&M requirements into account, and some minimum restrictions are required before optical optimization.

3.10 Conclusion

Parabolic-trough CSP systems have emerged as a promising technology for large-scale solar energy generation. By using curved mirrors to focus sunlight onto a receiver, they efficiently convert solar energy into thermal energy, which can then be used to drive turbines and generate electricity. These systems are highly scalable and have proven effective in both standalone and hybrid configurations with other renewable energy sources, enhancing grid stability.

The ability to integrate TES with CSP plants is a significant advantage, allowing for energy dispatchability even when the sun is not shining. Recent advances in materials, control systems, and HTFs have improved the efficiency and reduced the cost of CSP systems. Nevertheless, challenges remain, particularly in reducing the initial capital investment and ensuring reliable performance in varied climates.

In conclusion, parabolic-trough CSP systems are well positioned to play a critical role in the transition toward sustainable energy. Continued technological advancements and supportive policy frameworks are needed to overcome the remaining challenges, allowing CSP to contribute significantly to global energy needs, especially in sunny regions.

References

1 Ahmad, A., Prakash, O., Kausher, R. et al. (2024). Parabolic trough solar collectors: a sustainable and efficient energy source. *Materials Science for Energy Technologies* 7: 99–106.

2 Khan Faheem, M., Jizhan, L., Akram, M.W. et al. (2024). Design optimization, fabrication, and performance evaluation of solar parabolic trough collector for domestic applications. *Energy Sources, Part A: Recovery, Utilization, and Environmental Effects* 46 (1): 13896–13915.

3 Simeu, C.T., Njomo, D., Chara-Dackou, V.S., and Babikir, M.H. (2024). Comparative study of thermal performance improvement of parabolic trough solar collector using different heat transfer fluids, economic and environmental analysis of power generation. *Results in Engineering* 23: 102353.

4 Wang, Q., Yao, Y., Shen, Y. et al. (2024). A mutually beneficial system incorporating parabolic trough concentrating solar power system with photovoltaics: a comprehensive techno-economic analysis. *Applied Energy* 360: 122834.

5 Moya, E.Z. (2012). Parabolic-trough concentrating solar power (CSP) systems. In: *Concentrating Solar Power Technology*. (ed. K. Lovegrove and W. Stein), 197–239, Woodhead Publishing.

6 Babaei, A., Aminian, E., and Saffari, H. (2024). Enhanced heat transfer performance in a parabolic trough solar collector: a CFD investigation of silver and copper nanofluids utilizing porous medium. *Proceedings of the Institution of Mechanical Engineers, Part E: Journal of Process Mechanical Engineering* https://doi.org/10.1177/09544089241289072.

7 Al-Oran, O., Shaban, N.A., Manna, R. et al. (2024). Performance study of parabolic trough solar collector using hybrid nanofluids under Jordanian weather conditions. *Journal of Thermal Analysis and Calorimetry* 149 (9): 3981–3998.

8 Marušić, A. and Lončar, D. (2024). Different electric thermal energy storage configurations integrated with parabolic trough concentrating solar power plant. *Journal of Energy Storage* 94: 112472.

9 Jabbar, H.A., Alwan, K.J., Hachim, D.M. et al. (2024). Comparative assessment of thermal oils and water as working fluids in parabolic trough collectors for enhanced solar power generation. *Engineering Research Express* 6 (3): 035514.

10 Mohamad, K. (2024). Exploring the influence of optical and thermal parameters on the effectiveness of parabolic trough collector receiver units. *Clean Energy* 8 (4): 15–33.

11 Furlan, G. and You, F. (2024). Robust design of hybrid solar power systems: sustainable integration of concentrated solar power and photovoltaic technologies. *Advances in Applied Energy* 13: 100164.

12 Mohanavel, V., Logesh, K., Sharma, P. et al. (2024). Silicon porous disc featured parabolic trough collector functioned with paraffin: thermal performance study. *Applied Thermal Engineering* 257: 124392.

13 González-Mora, E. and Durán-García, M.D. (2024). Assessing parabolic trough collectors and linear Fresnel reflectors direct steam generation solar power plants in Northwest México. *Renewable Energy* 228: 120375.

14 Lüpfert, E., Geyer, M., Schiel, W. et al. (2001). Eurotrough design issues and prototype testing at PSA. *The American Society of Mechanical Engineers* 149: 387–391.

15 Mdallal, A., Haridy, S., Mahmoud, M. et al. (2024). Modelling and optimization of concentrated solar power using response surface methodology: a comparative study of air, water, and hybrid cooling techniques. *Energy Conversion and Management* 319: 118915.

16 Alaidaros, A.M. and AlZahrani, A.A. (2024). Thermal performance of parabolic trough integrated with thermal energy storage using carbon dioxide, molten salt, and oil. *Journal of Energy Storage* 78: 110084.

17 Khan, M.I., Asfand, F. and Al-Ghamdi, S.G. (2023). Progress in research and technological advancements of commercial concentrated solar thermal power plants. *Solar Energy* 249: 183–226.

4

Hybrid PV–CSP Systems

The massive deployment of photovoltaic (PV) electricity is increasingly challenged by the lack of dispatchability of this technology, despite the extremely low cost of PV energy in 2020 (recent announcements of PV power plant manufacturers relate to projects characterized by resale prices significantly below 0.03 EUR/kWh). Today, it is obvious that being unable to match the PV power generation with power demand leads to many power grid management issues. This is particularly evident in the challenges in managing output (solar power is injected into the power grid when sunlight permits, replacing the "dispatchable" power sources of nuclear, coal, and gas). Operators may be in an overproduction situation when PV energy production accounts for a sizeable portion of the energy demand (production and consumption balance must always be maintained to ensure the stability of the power grid), which may result in negative resale prices on the power markets. This frequently happens in Germany, where PV solar and wind power penetration rates have exceeded 30% of the nation's overall energy consumption. To meet the demand for electricity in the absence of sunlight (or wind, for wind power), the enormous deployment of variable renewable energy sources also necessitates the maintenance of large dispatchable power sources. There are various kinds of these sources: hydroelectric power, gas power plants, coal power plants, and nuclear power plants (some nations, like France, have load-following technology). This conundrum results from the necessity of maintaining CO_2-emitting technologies to make up for the intermittent nature of the solar resource: extensive deployment of PV power plants will not always be followed by a significant reduction in CO_2 emissions. It goes without saying that the creation of energy storage options is a promising means of overcoming the problem of solar intermittency, and numerous technologies are currently the subject of significant research and development efforts [1, 2]. Batteries, capacitors, flywheels, compressed air storage systems, pumped energy transfer stations, and hydrogen storage are a few examples worth mentioning. As of now, none of these technologies appears to be able to adequately address the enormous storage

Concentrated Solar Power Systems, First Edition. Bellamkonda Pragathi and D. P. Kothari.

requirements posed by the widespread use of variable renewable energy sources. On the one hand, most of these alternatives are currently unable to compete with conventional sources due to their high costs. On the other hand, numerous issues with material supply, recycling, and pollution are barriers to the widespread use of these different storage systems. Despite the fact that concentrated solar power is currently more expensive than PV, it still has one key advantage over PV: the ability to store thermal energy. Thermal energy storage, in contrast to many other energy storage methods, is a cheap and effective method that relies on readily available and low-polluting materials. Figure 4.1 crudely summarizes the appeal of hybrid strategies and shows how the major solar technologies are positioned on a cost-dispatchability diagram (dispatchability refers to a power production technology's potential to satisfy demand at a specific time). At the turn of the century in 2020, solar energy is unquestionably not very expensive, but it is also not very dispatchable. The inclusion of batteries (electrochemical storage) significantly improves the dispatchability of PV power plants; however, at the current cost of batteries, this increase in dispatchability is only marginal. While solar power plants offer lower electricity prices than concentrating solar power (CSP) plants, the latter have the advantage of better dispatchability. The recent development of hybrid PV–CSP technology demonstrates a desire to combine two advantages that are unique to each of these industries: the ease of use and

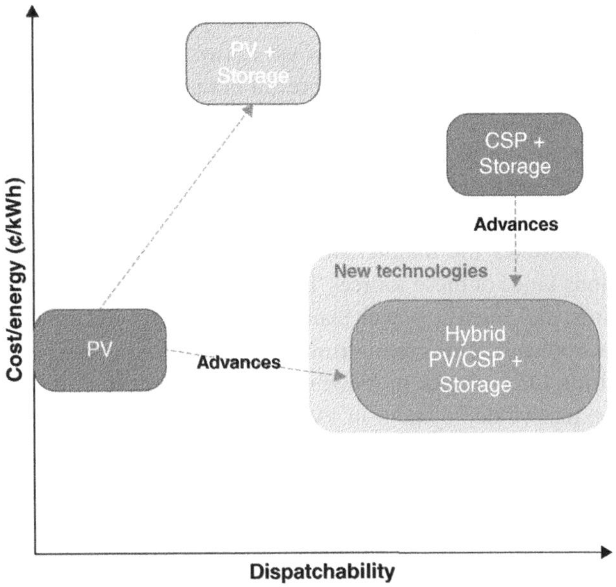

Figure 4.1 Cost and dispatchability comparison between PV technologies (with and without storage), CSP systems, and hybrid PV/CSP systems.

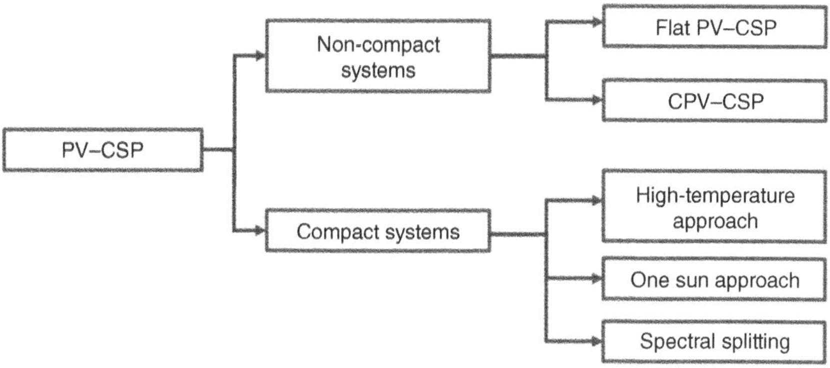

Figure 4.2 Main families of hybrid PV–CSP systems.

low cost of PV conversion technology as well as the ease of use and low cost of thermal storage technology found in CSP plants [3, 4].

4.1 Hybrid Strategies

There are various hybrid PV–CSP techniques in use today, and they primarily fall into two broad families shown in Figure 4.2.

1) "Noncompact" strategies, where PV and CSP plants are physically separated and run separately from one another.
2) The so-called compact strategies, which depend on combining the two technologies in a single power plant. It was decided to focus the majority of this chapter on the latter family of systems because of the various scientific and technological challenges that have been executed on compact hybrid systems, the wealth of literature that has been written about this family of systems over the years, and all of these factors.

4.2 Noncompact Hybrid Strategies

"Noncompact" hybrid systems are solar power plants using PV and CSP technologies that are constructed near to one another but run separately. With such systems, the solar resource is utilized to its fullest potential, with PV often used during the day and CSP used more frequently during cloudy or nighttime conditions (depending on the selected resale strategy). When compared to conventional PV or CSP plants, the use of noncompact hybrid systems significantly raises solar power plants' capacity factors. It's important to keep in mind that a power plant's capacity factor is the ratio between the power it produced over the course of the year and the power it would have produced had it been operated at rated power 100% of the time. Given the inherent flexibility of the technology, noncompact hybridization strategies also provide high levels of integration flexibility. For example, because the PV and CSP plants are constructed on separate (yet adjacent) sites, the PV power plant's characteristics have no bearing on the latter's performance. However, in the case of compact hybrid systems, the integration of these two technologies into a single power plant is unquestionably more difficult and presents real scientific and technological difficulties. The use of established, trustworthy, and tested solar technologies is a necessity for non-compact hybrid power plants, and their practical implementation is already taken into consideration, just as with other hybrid power systems. The hybrid PV/CSP plant project in Copiapó, *Chile,* is depicted in Figure 4.3 (IEEE Spectrum, n.d.) by an artist [5].

Figure 4.3 An artist's view of the noncompact hybrid PV–CSP plant of Copiapó (Chile). Source: Solar Reserve. For a color version of this figure, see www.iste.co.uk/flamant/energy.zip.

4.3 Compact Hybrid Strategies

The "compact" hybrid techniques basically belong to two sizable subfamilies based on different physical principles: Hybrid systems that utilize sun photons in diverse ways: these systems must include an optical component that directs photons to the best conversion system based on their characteristic wavelength. This family of systems can be separated into many subfamilies, such as the "one-sun" method and spectrum division, depending on the type of optical component used (PV cell, filter, and mirror).

4.3.1 High-Temperature Approach

A concentrating PV module serves as the thermal receiver for the high-temperature hybrid technique, also referred to as "PV topping" (Figure 4.4), which uses an integrated PV–CSP receiver. The thermal converter receives the heat that was collected by the module and either stores it for later use or transforms it into electricity. Because of this, the PV module must operate at a very high temperature, which is unquestionably the method's biggest scientific and technological challenge [6].

Figure 4.4 shows a "high-temperature" integrated PV–CSP receiver schematically. A high-temperature PV module that serves as both a thermal collector and a concentrator of solar energy absorbs the solar radiation. While some of the incident solar energy is transformed directly into electricity, the remainder is transported to the thermal collector as heat. Visit www.iste.co.uk/flamant/energy.zip to get a colored version of this figure.

In the PV community throughout the past few decades, the conversion of the waste heat produced by the PV cells was a topic of much research and created many

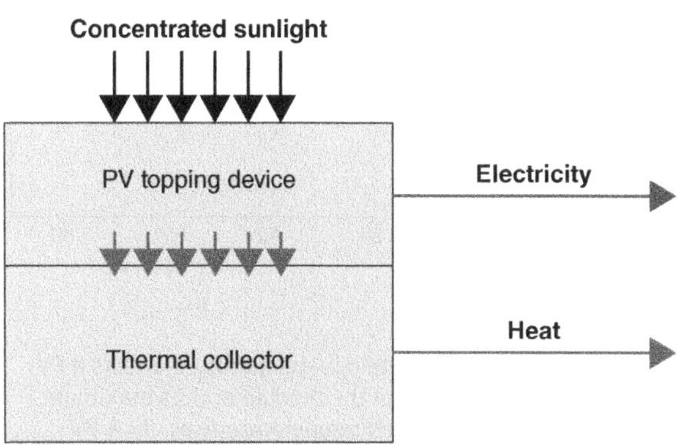

Figure 4.4 PV topping.

hopes. Today's research on the impact of temperature on PV cell performance mostly focuses on two separate phenomena [7, 8]:

- The increase in intrinsic carrier density, which is the concentration of charge carriers—electrons or holes—ejected from the valence band toward the conduction band or from the conduction band to the valence band under the sole effect of temperature, leads to a drop in the open-circuit voltage V_{oc} of the cell.
- The decrease in energy bandgap of semiconductor materials used in PV cells yields a slight increase in the photogenerated current I_{sc}. Overall, an increase in temperature results in a drop in the conversion efficiency of PV cells because the negative effect of temperature on the open-circuit voltage is more substantial in terms of its amplitude than the positive effect of temperature on the photogenerated current. This phenomenon is depicted in Figure 4.5: when a PV cell's temperature increases, it becomes substantially harder to extract electric power from it. The high-temperature hybrid approach's philosophy, which calls for typical PV module temperatures of several hundred degrees, and the phenomenon's amplitude appear irreconcilable.

In the past 10 years, a significant amount of research has been conducted on the temperature dependence of concentrating PV cells, which has resulted in a paradigm shift: solar flux concentration can significantly increase the temperature dependence of PV cells and, consequently, their ability to efficiently convert sunlight at high temperatures. This significant conclusion is explained by the fact that, as shown by the following equation, the open-circuit voltage V_{oc} of PV cells

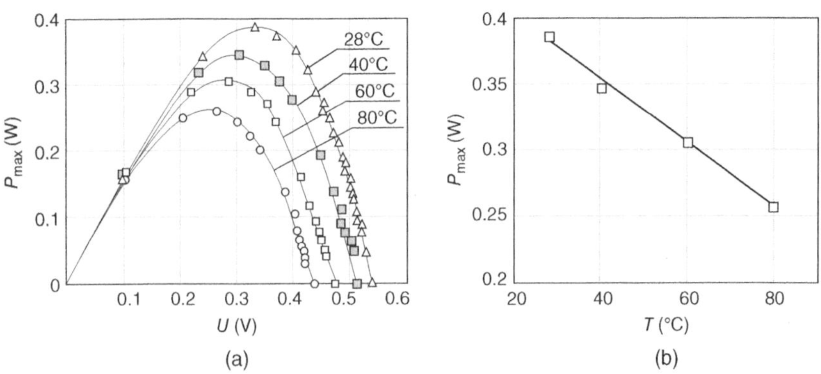

Figure 4.5 Power–voltage graphs (a) that demonstrate how temperature affects a PV module's electric properties The open-circuit voltage of the module and its maximum electric power both noticeably decrease as the module's temperature rises. (b) A PV module's maximum electric output as a function of temperature. Source: Adapted from Radziemska [9].

exhibits a dual dependence on temperature and the light power density to which the cells are exposed [10, 11].

The I_0 is the diode current (the current associated with the recombination of charge carriers in the cell) and I_{ph} is the current photogenerated by the cell (the current associated with the absorption of incident solar photons). T and X are the cell temperature and solar concentration factor, represented in suns, respectively (where 1 "sun" is equal to 1,000 W/m^2 of solar power density). While exposure to concentrated solar flux is accompanied by an increase in the photogenerated current, which varies proportionally to the absorbed solar power density (in practice, and as indicated above, there is also a positive effect of temperature on the photogenerated current, as is seen above), the increase in cell temperature causes an increase in the recombination current I_0. This is primarily because of the increase in intrinsic carrier concentration.

The temperature coefficient associated with a particular electric parameter of a solar cell, which specifies how a 1°C temperature increase affects the numerical value of the related parameter, is typically used to quantify the temperature dependency of that particular electric parameter. The temperature sensitivity clearly decreases with solar concentration, as shown by the evolution of the temperature coefficient of V_{oc} as a function of the solar concentration factor in Figure 4.6. In other words, because this particular cell is exposed to larger

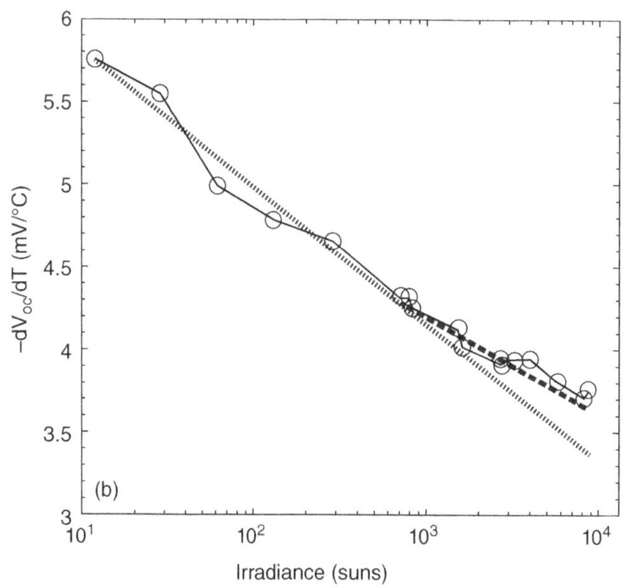

Figure 4.6 Temperature coefficient of V_{oc} as a function of the solar concentration ratio, for an InGaP/InGaAs/Ge multijunction solar cell. Source: Adapted from Braun et al. [12].

solar energy, the drop in the open-circuit voltage associated with cell heating is much smaller. Because the PV cells incorporated in the receiver are exposed to sufficiently high levels of solar concentration in light of these latest findings, the practical implementation of high-temperature hybrid systems can therefore be taken into consideration. In the case of high-temperature hybrid receivers, Figure 4.7 depicts the various energy fluxes (in percentage of the incident solar power) involved [14]. This is done while taking into account double-junction PV cells (composed of a stack of two pn junctions optimized to provide the

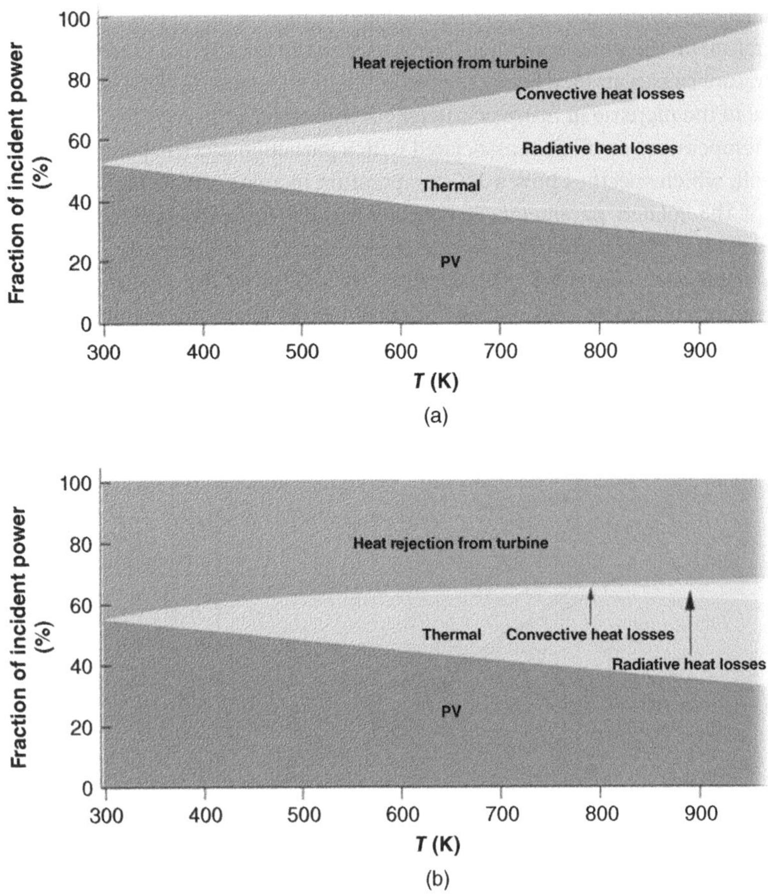

Figure 4.7 PV, thermal, and hybrid conversion efficiencies as a function of the operating temperature and the associated heat losses for cells operating at the radiative limit under a concentration of 100 suns (a) and at the radiative limit under 1,000 suns (b). Source: Vossier et al. [13]/with permission of Elsevier.

best solar energy conversion for a given temperature and concentration ratio). For concentration ratios of 100 suns (a) and 1,000 suns (b), the various energy fluxes connected to the high-temperature hybrid receiver are shown as a function of the receiver temperature. Finally, it is assumed that the turbine's conversion efficiency is equal to 2/3 of the Carnot efficiency. The hybrid system's electric power output, which is calculated as the sum of PV and thermal power, changes greatly depending on the temperature of the receiver and the solar concentration ratio to which it is exposed. As would be expected, as the hybrid receiver's temperature rises, the power produced by the PV module declines while the power produced by the turbine increases. A higher solar flux concentration ratio is advantageous on two fronts: (1) it causes the PV module's temperature sensitivity to be lower, increasing its conversion efficiency (and, consequently, its power) for rising temperatures; and (2) it significantly lowers (radiative and convection) heat losses, significantly increasing the electric power produced by the turbine for rising temperatures. It is critical to emphasize that, in addition to having a positive impact on maximum electric power, an increase in solar concentration ratio also significantly widens the range of temperatures for which the system performs at its best and makes it possible to more precisely control the equilibrium of the powers produced by each converter [15, 16].

The creation of solar cells that are effective at temperatures far higher than the ordinary operating temperatures of regular PV modules is, without a doubt, the biggest technological barrier to this strategy. Nevertheless, significant advancements were made during the past five years, particularly at NREL and Arizona State University, where PV cell topologies with acceptable performances at 400 and 600°C were produced. The development of three crucial high-temperature PV cell properties—open-circuit voltage V_{oc}, fill factor (FF), and conversion efficiency—is shown in Figure 4.8. Solar concentration enables a large gain in cell performances at a given temperature even if a distinct drop in the value of these three parameters can be seen with rising temperatures. These initial findings demonstrate how sun concentration can be used to create effective high-temperature hybrid systems. The development of effective high-temperature hybrid systems necessitates a better understanding of the physical mechanisms at play as well as an adaptation of PV cell architecture to extreme conditions of both illumination and temperature. It is important to note that the physics of PV cells at very high temperatures is still largely unknown [15, 18].

The creation of integrated receivers that operate at much lower temperatures offers an alternative to high-temperature compact hybrid systems. The development of such systems has been the focus of extensive research in recent years, particularly CPV-ORC (Organic Rankine Cycle) systems, which use Rankine organic cycle-based turbines and are characterized by temperatures that barely

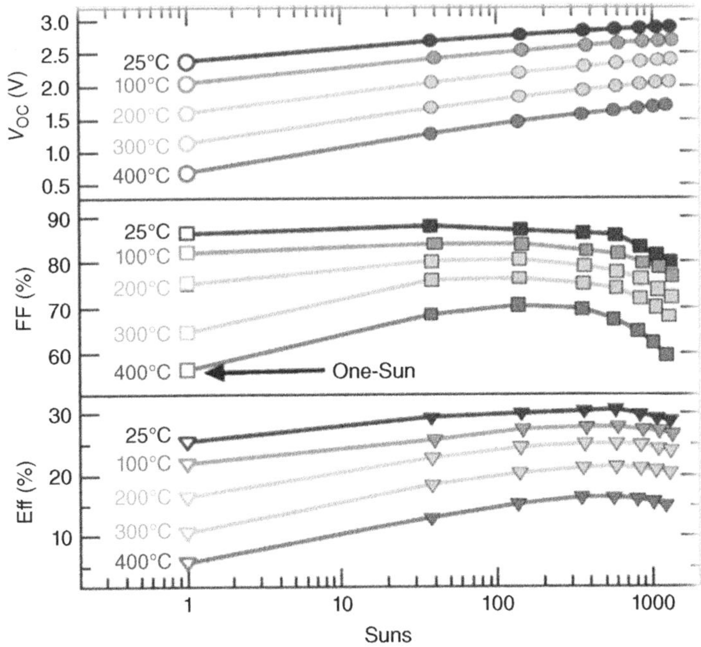

Figure 4.8 Open-circuit voltage (V_{oc}), fill factor (FF), and conversion efficiency of a GaInP/GaInAs tandem cell as a function of solar concentration and for temperatures ranging between 25 and 400°C. Source: Adapted from Perl et al. [17].

reach 100°C. The limitations on PV cells are greatly loosened by such a temperature level (because they can now operate at temperatures that are much closer to those found in traditional PV or CPV systems). The relevance and economic practicality of this approach, however, are seriously questioned by the ORC turbines' low efficiency over this temperature range.

4.3.2 Spectral Splitting

In that they are physically isolated from one another, the final two hybrid techniques described in this section differ from the high-temperature method. Depending on their wavelength, the incident solar photons are directed to one of these two converters or the other (high-energy photons are ideally absorbed by the thermal receiver, while low-energy photons are preferentially absorbed by the PV cells). One-sun approach and spectral separation are two unique subfamilies that can be distinguished by the layout of these two conversion systems and the type of optical component utilized for spectral separation [19].

4.3.2.1 PV One-Sun Approach

Using PV cells as both an optical reflector and a PV converter is known as the "PV one-sun" method. In actuality, a concentrator's surface is covered with PV cells that incorporate a mirror on the back (or alternatively, a stack of selective layers on the front) to allow PV cells to absorb the highest-energy photons. Photons with energies below the PV cell's energy bandgap are reflected in the direction of the CSP receiver. This method has several benefits in terms of PV cell operating conditions because it can actually convert some diffuse solar energy, as opposed to solar concentration systems that can only use the direct portion of radiation. Achieving high PV efficiencies is made possible by the characteristic working temperatures, which are much lower than in high-temperature systems. Figure 4.9 shows a schematic of a hybrid "PV one-sun" system with a parabolic trough geometry. The UV, mid-IR, and a significant portion of the visible spectrum are reflected to the thermal receiver, while the red and near-infrared (IR) components of the solar spectrum are transmitted through a film covering the solar cells. It is thus possible to precisely optimize the sun spectrum areas sent to each of the receivers, depending on the efficiency with which they will be converted, using an appropriate optical engineering of the selective layer. According to the "PV-mirror" concept's creators, who are associated with the University of Arizona, this technology allows for an increase in solar-to-electricity conversion efficiency of about 15% when compared to a traditional PV system [21].

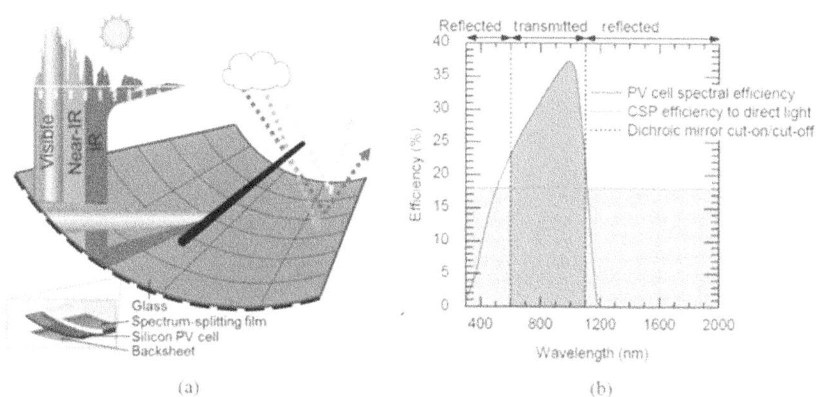

Figure 4.9 Schematic representation of the hybrid "PV-mirror" system developed at Arizona State University (a) and diagram representing spectral efficiencies of PV cells and CSP converter, (b) and the ranges of wavelengths used by each converter. Source: Fisher et al. [20]/AIP Publishing.

4.3.2.2 Strategies Based on the Spectral Separation of Light

The employment of an optical component that is integrated into the system and enables a spectrum separation of the incident light constitutes a second variant of the hybrid technique that involves spectral splitting of light. This method's theory is quite similar to the "PV one-sun" method that was previously discussed, with the distinction that the optical component used for spectral separation is not also a solar converter. This approach, which is similar to the "PV one-sun" approach, enables better control over the operating conditions on each of the converters (it's crucial to note that, unlike the "PV one-sun" approach, this system theoretically permits the regulation of the solar power density on the PV cells). An example of a hybrid solar power plant based on this methodology is shown in Figure 4.10. A spectrum splitter is put on the optical path of the light after the solar flux is first concentrated using a heliostat field. The photons that have an energy greater than the PV receiver's integrated cells' energy bandgap pass through the filter and go on to the latter, located at the top of the tower. The thermal receiver, located at the lower section, receives the low-energy photons that are reflected. With this design, the PV receiver can take advantage of the advantages of solar concentration while the PV and CSP conversions can be decoupled [15, 23].

Reduce the angular dependence of spectral separation systems; increase the transmission/reflection efficiency of the filters, which should ideally be close to a step function with values of transmissivity and reflectivity abruptly varying between 0% and 100% around a defined cutoff wavelength. For concentrating systems, which are typically characterized by a wide distribution of photons, the angular response of filters does actually have the tendency to change significantly with the incidence angle of photons.

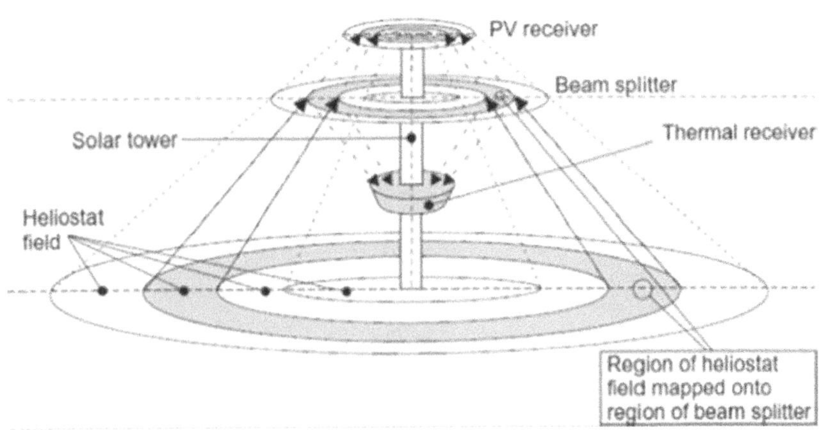

Figure 4.10 Example of hybrid PV–CSP plant based on the spectral separation of solar radiation. Source: Imenes et al. [22]/with permission of Elsevier.

4.3.3 Performance-Based Comparison of the Main Hybrid Strategies

The evaluation of a hybrid system's performance is more complicated than it initially appears. It's crucial to keep in mind that these systems must simultaneously guarantee the highest solar-to-electricity conversion efficiency and a good balance between PV electricity and thermal electricity. This challenge is demonstrated in Figure 4.11: As a function of the energy bandgap of the PV cell used, the graphic on the left shows the theoretical maximum conversion efficiency of "high-temperature" hybrid systems, hybrid systems based on the spectral splitting of light, and hybrid systems using the "PV one-sun" approach. The equilibrium between PV generation and CSP generation is shown in the graphic on the right, also as a function of the energy bandgap of the PV cell. If the PV fraction is 1, then all of the system's electricity comes from PV sources. Conversely, a number of 0 denotes that the thermal converter is the sole source of electricity.

A number of significant inferences can be made from Figure 4.11. Although they are observed for noticeably different energy bandgaps (1.1 eV for the "high-temperature" approach and 1.5 eV for the "PV one-sun" approach), the "PV one-sun" and "high-temperature" approaches, among the three hybrid PV/CSP strategies represented, have very similar global conversion efficiencies. Contrarily, the final method—using the spectrum splitting of light—has a substantially lower overall conversion efficiency. The "high-temperature" approach is characterized by a balanced distribution between PV and CSP over a wide range of bandgaps (the PV fraction ranging between 0.4 and 0.6 for values of E_g between 0.4 and 2 eV). This equilibrium between PV and CSP, however, varies particularly as a function of the strategy under consideration. The two methods that rely on

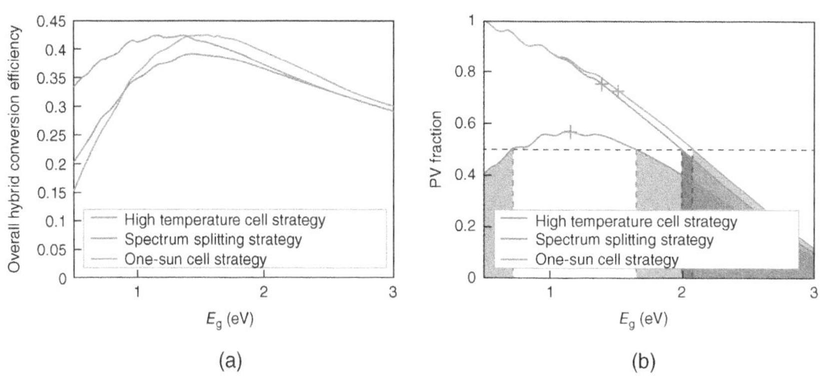

Figure 4.11 Global conversion efficiency (a) and PV fraction (b) for three compact hybrid strategies, as a function of the energy bandgap of the PV cell, and for PV cells operating at the radiative limit. Source: Vossier et al. [13]/The Royal Society of Chemistry.

spectral selection of input photons, however, exhibit a significant imbalance in favor of PV over a wide range of bandgaps. Additionally, despite the fact that the "PV one-sun" approach has a high global conversion efficiency, the strong imbalance in favor of PV indicates that this approach is unlikely to be able to guarantee sufficient CSP contribution, limiting the system's ability to generate electricity outside of sunshine periods.

4.4 Hybrid PV–TS Systems

It is crucial to keep in mind that the primary driver behind hybrid PV–CSP systems is the dramatic decrease in solar energy costs achieved by the integration of two separate, low-cost technologies: PV solar power plants and thermal energy storage. By directly connecting these two elements, a radical expansion of this idea might be realized without the need for the CSP plant that is typically used for thermal energy storage. Figure 4.12 illustrates this idea conceptually. A typical PV power

Figure 4.12 Illustration of the PV + thermal storage + steam turbine concept [24]. For a color version of this figure, see www.iste.co.uk/flamant/energy.zip.

plant is connected to the grid, and any excess electricity is converted to heat by resistive heating and transmitted to thermal storage materials (often molten salts). The heat that has been stored can then be used to power a turbine at night or during cloudy periods. This strategy has a major flaw that prevents its practical application, despite the fact that it initially appears appealing due to its relative simplicity (since it does not necessitate the construction of a conventional CSP plant in addition to the PV power plant).

The Carnot efficiency, which is normally around 45%, actually determines the efficiency of the heat into electricity conversion. Furthermore, given the number of conversion processes involved and their relative inefficiency (20% efficiency for solar power-to-PV electricity conversion and 45% efficiency for thermal power-to-electricity conversion), the overall efficiency of solar power-to-electricity conversion of this approach is particularly low. However, compared to solutions with significantly higher performances but requiring expensive components or technologies, this solution may be economically attractive given the very low cost of PV electricity generation at the turn of the century and the very low cost of heat thermal storage. Additionally, this approach offers greater flexibility compared to conventional hybrid systems (better use of the diffuse solar resource, which is underutilized or not used at all by hybrid systems, the possibility of fully decoupling the PV field from the thermal energy storage and the turbine, possibility of using this strategy in latitudes where concentrating solar power plants are less feasible to install, etc.). Now, this idea is still being explored and argued about, and its economic advantage is still unclear.

4.5 Innovative Hybrid Systems

In recent years, novel ideas have been created to maximize collecting and conversion while enhancing the dispatchability of solar energy conversion systems. These ideas make use of the "exotic" features and mechanisms of the materials. An incomplete review of some recently proposed solutions is provided in this section. They all rely on the employment of a high-temperature storage fluid, which enables the timing of power generation [25].

4.5.1 Mixed Hybrid Systems

The numerous compact hybrid systems discussed in this chapter can be roughly divided into two major families that can be characterized by the types of physical mechanisms used to harvest, divide, and steer the solar resource toward the respective PV and CSP receivers. Due to the physical separation of the two converters, the systems based on spectral splitting of light provide a degree of implementation freedom. However, it is challenging to use PV cells in these systems to their full potential. In fact, the use of very high-efficiency PV cells is prohibited due to the wide cell surfaces that are necessary and the low power density to which these cells are exposed. In addition to the ability to expose the PV cells to concentrated solar flux, which has the benefit of improving conversion

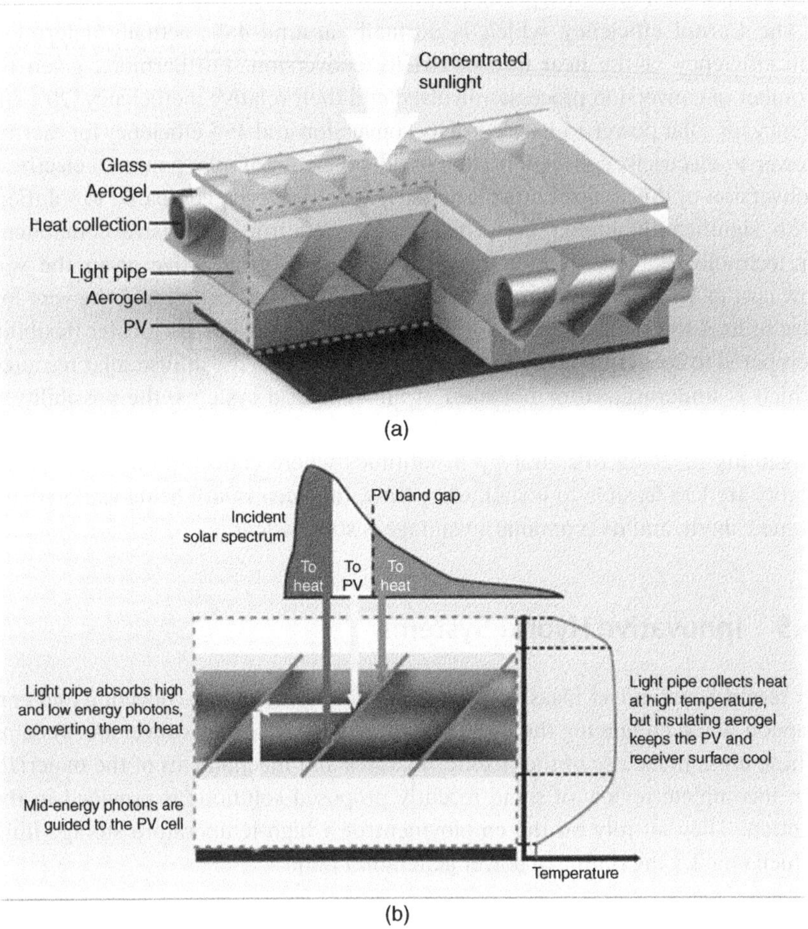

Figure 4.13 Schematic representation of the HEATS concept. (a) The module consists of a stack of several elements, notably a glass protection, an aerogel layer acting as a thermal insulator, spectrally selective light pipes, another aerogel layer, and, finally, PV cells. The light pipes contain parallel fins that are coated with a selective coating. These fins are attached to the light pipes, to which they transmit the harvested heat. (b) The latter is absorbed by the heat transfer fluid and is then carried to the thermal storage system. Source: Weinstein et al. [26]/with permission of Elsevier.

efficiency and significantly reducing the required cell surface, high-temperature systems also offer the advantage of receiver compactness. However, the development of this latter strategy is undoubtedly hampered by the high temperatures to which these cells must constantly be exposed. A potential answer is the creation of mixed hybrid receivers, which are based on the intersection of these two approaches and are conceptually comparable to the HEATS technology created by MIT and shown in Figure 4.13. The HEATS system shown in Figure 4.13 uses "light pipes" to absorb low-energy photons while transmitting the highest-energy photons to the PV cells on the module's back. Light pipes do not necessarily need to be in thermal contact with the PV cells, which can work at temperatures close to the ambient temperature, unlike a standard high-temperature receiver. Additionally, a thermal insulator that is transparent in the solar spectrum is coated on the module's external aperture, allowing for efficient light transmission to the "light pipes" and then to the PV cells while keeping the module's surface temperature as low as feasible. This offers each converter the best possible operating conditions while drastically lowering heat losses.

The use of a photoluminescent absorber is a novel idea recently proposed to adapt the spectral content of light to the absorption properties of solar converters. This physical mechanism relies on the absorption of solar radiation by materials that reemit photons over a limited range of wavelengths (Figure 4.14).

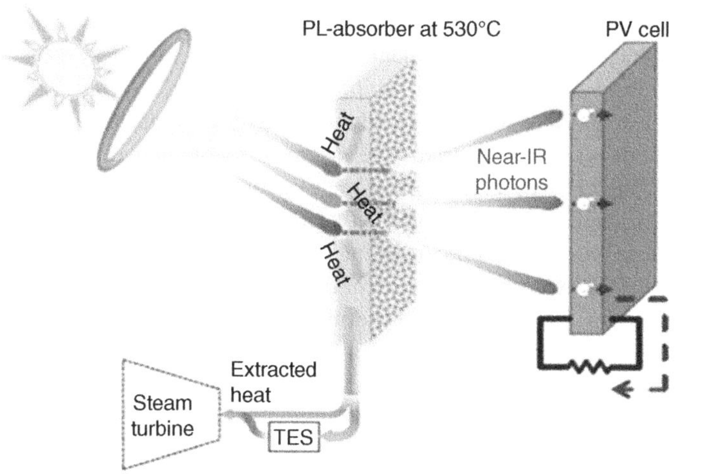

Figure 4.14 Illustration of LSC concept. The solar flux is concentrated on photoluminescent absorber. The radiation emitted by photoluminescence (whose spectrum is red shifted) is absorbed by a solar cell whose energy bandgap is adapted, while the residual heat (at 530°C) is stored to be injected in a turbine. Source: Haviv et al. [27]/American Chemical Society/CC BY 4.0.

4.5.2 Luminescent Solar Concentrators

It refers to the luminescence-based collecting and conversion of light, which should not be confused with this idea. In an LSC, solar radiation is focused on a photoluminescent receiver that absorbs high-energy photons and then emits mid-energy photons in the direction of PV cells positioned behind the absorber. The spectrum dispersion of photons received by solar cells is tailored to their absorption characteristics, minimizing thermalization losses and ensuring maximum PV conversion efficiency. A heat transfer fluid absorbs low-energy photons and the heat generated during the conversion of photons in the PL absorber before evacuating it. The thermal energy that has been obtained in this way is then either used immediately in a turbine or stored for later use. It is significant to remember that controlling the flow rate of the transfer fluid in the PL absorber enables adjustment of both its temperature and the percentage of photons either supplied to or absorbed by the PV cells.

4.5.3 Very High-Temperature Thermal Energy Storage Coupled with Photovoltaic Conversion

The majority of "hybrid" approaches for converting heat into electricity rely on the utilization of turbines. These have numerous benefits in that the technology is dependable, tested, and has great conversion efficiency. However, because the characteristic temperature of the body emitting the radiation is low or moderate, PV cells perform poorly when converting high-temperature heat sources, such as solar radiation, which is roughly equivalent to the radiation emitted by a black body at a temperature of 5,700 K. As a result, the preference for adopting one or the other of these methods depends greatly on the typical temperature of the energy source that needs to be converted. The vast majority of the hybrid systems discussed in this chapter rely on a thermal storage system that uses molten salts, which are typically heated to temperatures between 500 and 600°C. The employment of PV cells as a mechanism for converting heat into electricity can be taken into consideration if the typical temperature of the storage medium is noticeably higher. The systems for PV conversion of thermal energy at very high temperatures (TEGS-MPV) are based on this theory. In Figure 4.15, the workings of such a system are graphically illustrated. In the TEGS-MPV proposal, which was put up by an MIT team in 2019, silicon baths operating at extremely high temperatures serve as both thermal storage and radiation sources. A "cold" tank holding liquid silicon is maintained at 1,900°C (discharged state) temperature. Liquid silicon is pumped via pipes heated by heating resistors made of tungsten or graphite to charge the system, allowing for the achievement of a "hot" temperature of 2,400°C. The electricity produced by a conventional power plant (PV, wind, hydraulic, etc.) powers the pumping and heating systems. The "hot" pump pumps liquid silicon

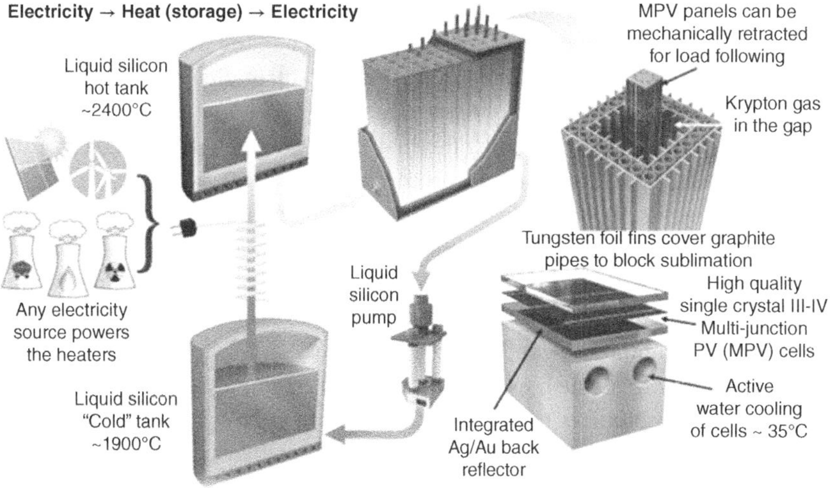

Figure 4.15 Schematic representation of TEGS-MPV concept: electricity is converted in the form of heat through a heating resistor, before being transferred to a liquid storage in the form of latent heat (1,900–2,400°C). Low-cost metallurgical-grade silicon is used as a storage material. When electricity must be generated, the liquid is pumped through a set of pipes, emitting radiation converted into electricity by multijunction cells located at the center of the set. Source: Amy et al. [28]/The Royal Society of Chemistry.

into the MPV cycle when the power plant operator wants to inject electricity into the grid. This cycle is made up of a number of graphite pipes that are inserted into a tungsten container, which serves as a barrier of protection between the graphite pipes and the PV module that is inserted at the assembly's center. The PV cells are exposed to photons from the interior tungsten wall, which acts as a photon emitter (much like an incandescent bulb), and some of these photons are converted into energy by the PV cells. Liquid silicon gradually cools as it travels through the network of graphite pipes before returning to the "cold" tank, where it can be used once more. This storage method may be effective and affordable, but it faces a number of clear scientific and technological obstacles, making its short-term application unlikely. However, it shows the wide range of storage options being researched today to make solar electricity an affordable, effective, and dispatchable alternative.

4.6 Conclusion

Hybrid photovoltaic–thermal solar (PV–TS) systems offer an efficient solution to maximize solar energy utilization by simultaneously generating electricity and thermal energy from the same installation. These systems improve the overall energy conversion efficiency by capturing and utilizing the waste heat generated by PV panels, which would otherwise be lost. By integrating thermal collectors with PV panels, hybrid PV–TS systems reduce the panel temperature, improving electrical efficiency while providing thermal energy for applications such as water heating, space heating, or industrial processes.

The key benefits of hybrid PV–TS systems include higher overall energy yields, better space utilization, and improved energy efficiency compared to standalone PV or thermal systems. Recent advancements in materials, design, and cooling technologies have further enhanced their performance, making them more viable for residential, commercial, and industrial use.

However, challenges such as higher initial costs, complexity in system design, and the need for optimized control strategies remain. Addressing these challenges through technological innovation and large-scale deployment can accelerate the adoption of hybrid PV–TS systems, contributing to sustainable energy solutions and reducing reliance on fossil fuels.

In conclusion, hybrid PV–TS systems represent a promising technology that can play a crucial role in the transition toward renewable energy by maximizing solar energy capture and utilization, thereby contributing to the overall sustainability and efficiency of energy systems.

References

1 Alam, K., Mostakim, M.A., Baki, A.A., and Hossen, M.S. (2024). Current trends in photovoltaic thermal (PVT) systems: a review of technologies and sustainable energy solutions. *Academic Journal on Business Administration, Innovation & Sustainability* 4 (04): 128–143.

2 Mortadi, M., El Fadar, A., and Achkari, O.B. (2024). Concentrating and non-concentrating photovoltaic thermal collectors: technologies, applications, exhaustive assessment and challenges. *Solar Energy Materials and Solar Cells* 271: 112858.

3 Dehghan, M., Vajedi, H., Rahgozar, S., and Karimi, N. (2024). Energy, economic, and environmental analysis of converging air-based photovoltaic-thermal (air/PV-T) systems: a yearly benchmarking. *Journal of Cleaner Production* 434: 139871.

4 Khan, M.S., Wang, J., Memon, A.A., and Muhammad, T. (2024). Investigating the enhanced cooling performance of ternary hybrid nanofluids in a three-dimensional annulus-type photovoltaic thermal system for sustainable energy efficiency. *Case Studies in Thermal Engineering* 60: 104700.

5 Alnakeeb, M.A., Salam, M.A.A., Hassab, M.A., and El-Maghlany, W.M. (2024). Numerical study of thermal and electrical performance of a new configuration of hybrid photovoltaic solar panel phase-change material cooling system. *Journal of Energy Storage* 97: 112945.

6 Tripty, T.A. and Nasrin, R. (2024). Efficiency upgrading of solar PVT finned hybrid system in Bangladesh: flow rate and temperature influences. *Heliyon* 10 (7): e28323.

7 Yildirim, M.A. and Cebula, A. (2024). A numerical and experimental analysis of a novel highly-efficient water-based PV/T system. *Energy* 289: 129875.

8 Arévalo, P., Ochoa-Correa, D., and Villa-Ávila, E. (2024). Advances in thermal energy storage systems for renewable energy: a review of recent developments. *Processes* 12 (9): 1844.

9 Radziemska, E. (2003). Thermal performance of Si and GaAs based solar cells and modules: a review. *Progress in Energy and Combustion Science* 29 (5): 407–424.

10 Li, R., Zhai, P., Li, J., and Liu, X. (2024). Experimental study and performance enhancement of micro heat pipe PV/T system. *Energy and Built Environment.* https://doi.org/10.1016/j.enbenv.2024.02.008.

11 Said, Z., Bellos, E., Ali, H.M. et al. (2025). Nanofluids, turbulators, and novel working fluids for heat transfer processes and energy applications: current status and prospective. *Applied Thermal Engineering* 258: 124478.

12 Braun, A., Hirsch, B., Vossier, A. et al. (2013). Temperature dynamics of multijunction concentrator solar cells up to ultra-high irradiance. *Progress in Photovoltaics: Research and Applications* 21 (2): 202–208.

13 Vossier, A., Zeitouny, J., Katz, E.A. et al. (2018). Performance bounds and perspective for hybrid solar photovoltaic/thermal electricity-generation strategies. *Sustainable Energy & Fuels* 2 (9): 2060–2067.

14 Uhanto, U., Yandri, E., Hilmi, E. et al. (2024). Predictive maintenance with machine learning: a comparative analysis of wind turbines and PV power plants. *Heca Journal of Applied Sciences* 2 (2): 87–98.

15 Kazem, H.A., Chaichan, M.T., Al-Waeli, A.H. et al. (2024). Energy enhancement of building-integrated photovoltaic/thermal systems: a systematic review. *Solar Compass* 12: 100093.

16 Cygan, D., Abbasi, H., Pondo, J. et al. (2024). *Hybrid Solar System* (No. DOE-GTI-AR0001102). Gas Technology Institute.

17 Perl, E.E., Simon, J., Friedman, D.J. et al. (2018). (Al)GaInP/GaAs tandem solar cells for power conversion at elevated temperature and high concentration. *IEEE Journal of Photovoltaics* 8 (2): 640–645.

18 Mdallal, A., Haridy, S., Mahmoud, M. et al. (2024). Modelling and optimization of concentrated solar power using response surface methodology:

a comparative study of air, water, and hybrid cooling techniques. *Energy Conversion and Management* 319: 118915.

19 Qudah, A., Almerbati, A., and Mokheimer, E.M. (2024). Novel approach for optimizing wind-PV hybrid system for RO desalination using differential evolution algorithm. *Energy Conversion and Management* 300: 117949.

20 Fisher, K., Yu, Z.J., Striling, R., and Holman, Z. (2017). PVMirrors: Hybrid PV/CSP collectors that enable lower LCOEs. *AIP Conference Proceedings* 1850 (1): 020004.

21 Başaran, K. and Koç, İ. (2024). PV/T system application for renewable heat and electric energy in buildings: performance and techno-economic analysis. *Isı Bilimi ve Tekniği Dergisi* 44 (2): 359–373.

22 Imenes, A.G., Buie, D., Mills, D.R. et al. (2006). A new strategy for improved spectral performance in solar power plants. *Solar Energy* 80 (10): 1263–1269.

23 Ghasempour, R., Salehi, M., and Fakouriyan, S. (2024). Comprehensive energy, exergy, and environmental evaluation of integrated heat pipe photovoltaic/thermal systems using multi-objective optimization. *Applied Thermal Engineering* 257: 124402.

24 Gordon, J.M., Fasquelle, T., Nadal, E., and Vossier, A. (2021). Providing large-scale electricity demand with photovoltaics and molten-salt storage. *Renewable and Sustainable Energy Reviews* 135: 110261.

25 Xu, X., Ma, T., Feng, X. et al. (2024). Simulation and optimal configuration of a combined photovoltaic-thermal and heat pump system with a hybrid energy storage. *Energy Conversion and Management* 317: 118851.

26 Weinstein, L.A., McEnaney, K., Strobach, E. et al. (2018). A hybrid electric and thermal solar receiver. *Joule* 2 (5): 962–975.

27 Haviv, S., Revivo, N., Kruger, N. et al. (2020). Luminescent solar power–PV/thermal hybrid electricity generation for cost-effective dispatchable solar energy. *ACS Applied Materials & Interfaces* 12 (32): 36040–36045.

28 Amy, C., Seyf, H.R., Steiner, M.A. et al. (2019). Thermal energy grid storage using multi-junction photovoltaics. *Energy & Environmental Science* 12 (1): 334–343.

5

Solar Fuels

5.1 Introduction to Solar Fuels

Carbonaceous materials can be converted into gaseous or liquid solar fuels using concentrated solar energy. These materials can be solid carbonaceous (coal, biomass, organic waste, etc.) or liquid or gaseous (oils, natural gas, etc.). The endothermal hydrocarbon cracking and reforming processes, as well as the pyrolysis and gasification of carbonaceous materials, are particularly relevant here (Figure 5.1). The use of concentrated solar energy to provide heat for these processes has a number of benefits, including [1, 2]

- Avoiding the combustion of fossil fuels to generate heat, which lowers the global consumption of hydrocarbons;
- Limiting the production of CO_2 and thereby reducing the impact of climate change;
- Conserving raw materials because they are no longer burned;
- Storing solar energy in the products after it has provided the enthalpy of reactions; producing cleaner fuels that can be used in other;
- Depending on the concentration ratios, emits heat at extremely high temperatures with very considerable radiative fluxes, enabling particular reaction pathways.

The performance of solar reactors is primarily determined by two metrics. The energy upgrade factor, which is sometimes represented by the letter U, is the proportion of energy (higher calorific value, or HCV), in the products to that in the reactants. The desired value is $U > 1$, which indicates that the products contain some solar energy.

The energy of the products is compared to the energy of the reactants plus the solar energy used to determine the efficiency of solar energy conversion into fuel. The use of solar energy is accounted for by this ratio, which is always less than 1, and if it is greater than 1, all solar energy used was stored in the products.

Concentrated Solar Power Systems, First Edition. Bellamkonda Pragathi and D. P. Kothari.

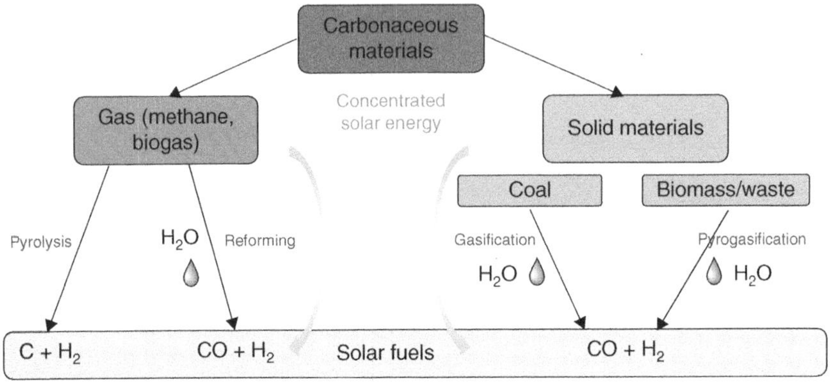

Figure 5.1 Main processes for manufacturing synthetic fuels from hydrocarbon resources. Source: Adapted from www.iste.co.uk/flamant/energy.zip.

5.2 Solar Cracking and Reforming of Hydrocarbons

Thermal dissociation, sometimes referred to as solar cracking, is the breakdown of a hydrocarbon as a result of temperature. The overall response can be expressed as: Because the dissociation reaction is endothermal, it needs an energy input that can be provided by concentrated sun radiation. A gaseous fuel (H2) and solid carbon in the form of nanoparticles (carbon black), the end products of which are primarily utilized industrially for the production of tires. For temperatures above 1,300 K, the reaction is complete at thermodynamic equilibrium, as shown in Figure 5.2. Kinetic factors should also be taken into account. Acetylene, one of the primary reaction intermediates, is one of the subproducts that can develop depending on the operating circumstances (temperature, residence time). Figure 5.3 shows how the reaction's energy balance is balanced. One mole of CH_4 has an energy content of 889 kJ (HCV). Following the molecule's solar dissociation, the products' energy value rises due to the reaction's enthalpy, which is provided by solar energy. Saving fossil fuels (and preventing CO_2 emissions as a result) is the motivation behind this solar energy response, which also serves to store solar energy [5, 6]. Solar energy accounts for 8% of the products' energy content ($U = 1.08$). Moreover, hydrogen and carbon both comprise 41% and 59% of the energy in products, respectively.

It is crucial to remember that this method can be used to change a fossil resource (methane) into a gas with no carbon emissions (hydrogen). Thus, this procedure is viewed as a stage in the transition from a fossil fuel-based economy to one that relies on the usage of renewable energy sources. In fact, thermal dissociation would be a way to keep using the extensively used gas infrastructures while preventing CO_2 emissions from methane burning, provided that the energy needed for the conversion process is renewable. From this angle, solar cracking has been studied for a very long time.

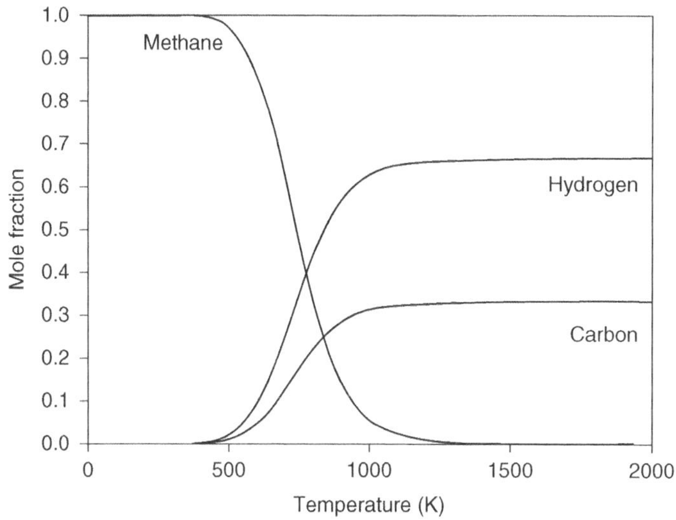

Figure 5.2 Simplified equilibrium diagram of the dissociation of CH_4 depending on the temperature. Source: Hirsch [3]/with permission of Elsevier.

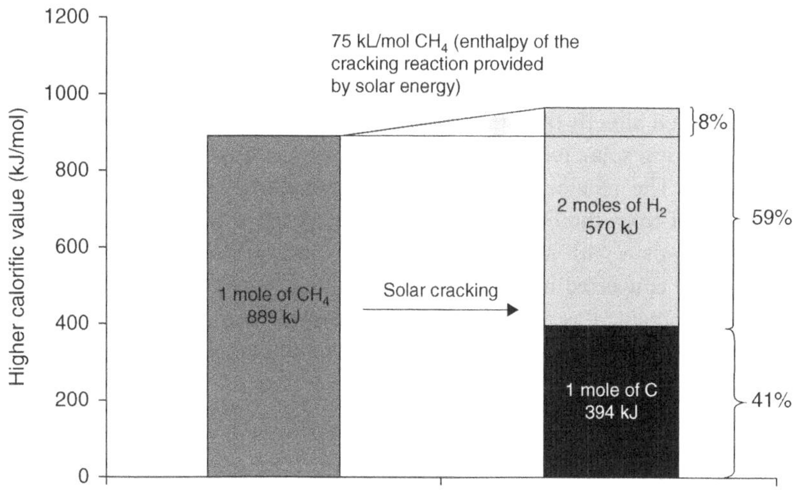

Figure 5.3 Energy balance of the solar cracking of methane. Source: Rodat et al. [4]/with permission of Elsevier.

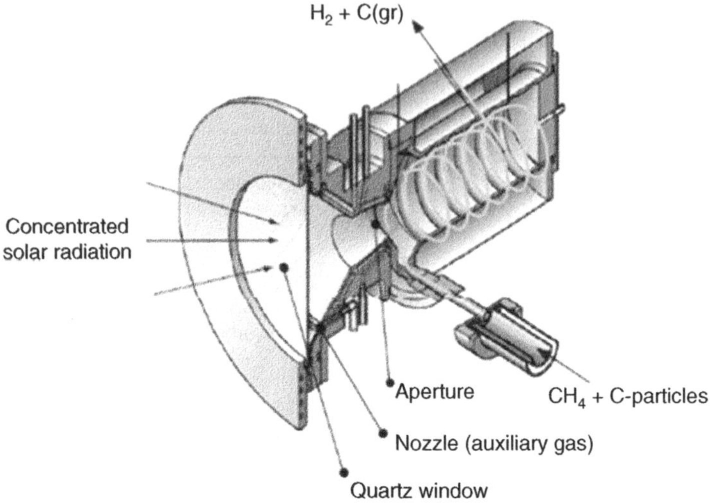

Figure 5.4 Direct heating solar reactor for methane cracking. Source: Hirsch [7]/with permission of Elsevier.

The initial publications were done in 1978 suggested conducting a review of recent publications. Numerous reactors were investigated and divided into two groups:

- **Direct heating reactors**: These reactors use a seeding of particles that act as radiative absorbers to provide a large transfer surface, which allows the reactant gas to be heated directly by sun radiation.
- A direct heating solar reactor was the focus of Hirsch and Steinfeld's work (Figure 5.4). The reactor is made up of a cylindrical compartment that is windowed off from the outside world. The methane is tangentially injected. Because it is sown with active coal particles, concentrated solar light may be absorbed and converted into heat for the reaction. Very considerable transfer surfaces are obtained as a result of the injection of 10 m particles. At 1,600 K and 5 kW power, methane could be converted into hydrogen with a maximum efficiency of 67% [6, 8].

5.3 Indirect Heating Reactors

In indirect heating reactors, the reactant gas does not directly absorb solar energy; instead, it is absorbed by an absorber's walls and released to the reactant gas via radiation, conduction, and convection. Particle seeding is not required, and there is no issue with particle deposit on the window, despite the fact that the heat

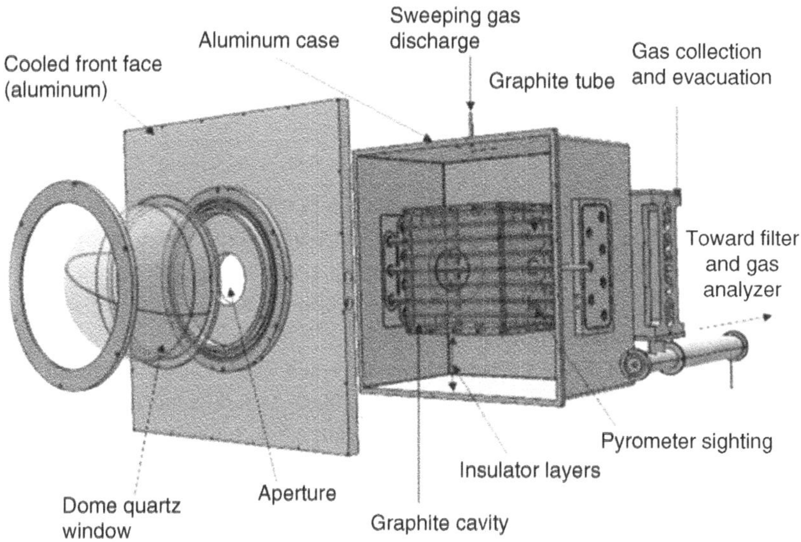

Figure 5.5 Indirect heating solar reactor for methane cracking. Source: Abanades [9]/with permission of Elsevier.

transfer surface is lower than that for direct heating reactors. At PROMES-CNRS, two indirect heating solar reactor prototypes with thermal outputs between 10 and 50 kW were created. To play the part of the black body, they use tubes that are placed inside a chamber (Figure 5.5). Graphite is the substance used, and at high temperatures, it has good optical, thermal, and mechanical qualities. The receiving cavity is isolated by a glass and swept by an inert gas to prevent the material from oxidizing. From a technological extrapolation standpoint, the lack of a window may be possible due to the usage of oxidation-resistant materials. Another constraint is the heat transfer to gas. A 50 kW reactor produces 200 g/h of H_2 (88% efficiency in H_2), 330 g/h of carbon black (49% efficiency in carbon), and 340 g/h of C_2H_2 with a thermal efficiency of 15% for 900 g/h of CH_4 injected (mole fraction of 50%, the remainder being argon) at 1,800 K. The most significant subproduct is acetylene. It can be utilized as is or its content can be diminished by lengthening the stay [10]. Metallic- or carbon-based catalysts can be utilized to restrict the reaction temperatures.

Other reactors that could be used with solar energy have been created. For instance, a packed-bed reactor was designed where methane is dissociated in a vector gas, such as hydrogen or nitrogen, heated by particles. Another approach involved a liquid metal reactor, using metals like copper or tin, where carbon is separated through density gradients, and heat is utilized to convert the gas into methane through bubbling, similar to the recovery of slags in steel baths. In fact, maintaining gas-phase cracking solar reactors is still difficult due to the control of carbon (and pyrolytic deposits on the walls). The reforming process described in the following section is brought on by the addition of water vapor to the system.

5.4 Solar Reforming of Natural Gas

In the process of reforming, methane reacts with either water or carbon dioxide to create synthetic gas, a mixture of hydrogen and carbon monoxide. At the moment, this method of producing hydrogen is the most popular. In contrast to steam reforming, the reaction of methane with carbon dioxide is known as dry reforming. The following responses are at work:

On metal catalysts of the nickel type, these extremely endothermic reactions are carried out at temperatures of the order of 800–900°C. Traditionally, methane is burned to create heat, but this process produces large CO_2 emissions. The process can reduce the amount of methane it uses by around 40% by adding solar thermal energy at high temperatures. Figure 5.6 shows the impact of temperature on

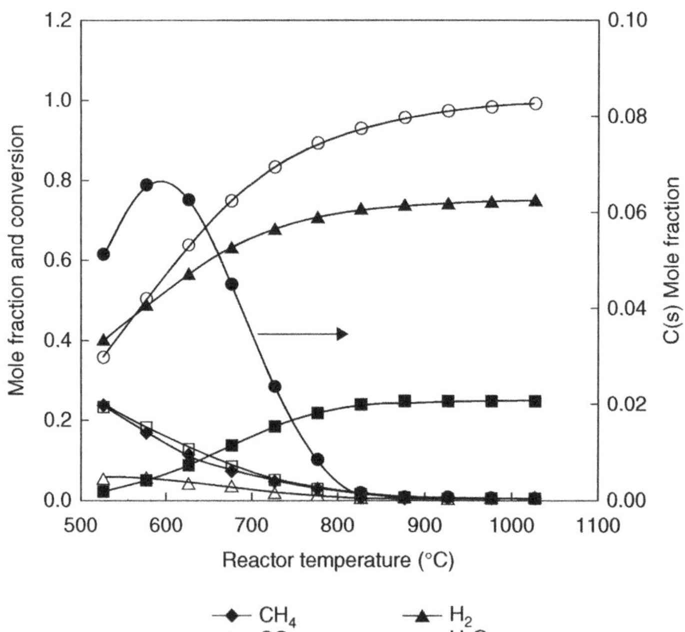

Figure 5.6 Diagram of equilibrium and conversion of methane during the steam reforming reaction (pressure 1 bar, stoichiometric ratio $CH_4/H_2O = 1$). Source: Seo et al. [11]/with permission of Elsevier.

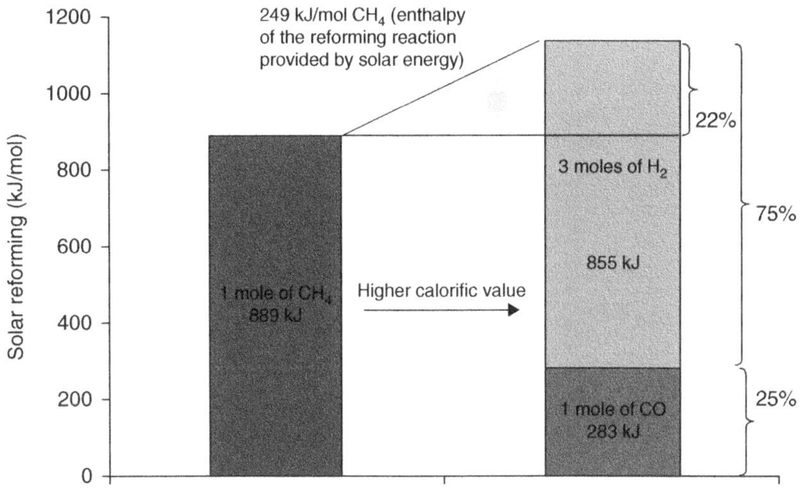

Figure 5.7 Energy balance of the solar steam reforming of methane. Source: Adapted from www.iste.co.uk/flamant/energy.zi.

the CH_4/H_2O equimolar mixture's composition at thermodynamic equilibrium. The only byproducts produced above temperatures of about 900°C are H_2 and CO, with an H_2/CO ratio of 3. Low pressures (Le Chatelier's principle) also help the reaction [12].

Reforming is typically followed by a stage of conversion through the exothermal reaction of the water-gas shift to transform carbon monoxide into hydrogen: According to: "From an energy standpoint, the steam reforming reaction is equivalent to the progression of the methane cracking reaction and subsequent carbon gasification."

Figure 5.7 depicts the reaction's energy balance. With 22% ($U = 1.22$) of the final energy coming from solar-derived products (compared to 8% for pyrolysis), it can be shown that the reaction is unquestionably more endothermal than the pyrolysis reaction. The remaining one-fourth is gained in the form of CO, leaving three-fourths of the total energy in the form of H_2.

5.4.1 State of the Art

Several literature studies have provided comprehensive overviews of the numerous works on the topic of solar reforming. There are three different kinds of reactors: those heated by an intermediary fluid, tubular reactors heated indirectly by the sun, and volumetric reactors heated directly by the sun. There has been research on a number of reformers heated by an intermediary fluid (sodium, air), which is heated by solar energy. For instance, the ASTERIX (Advanced Steam

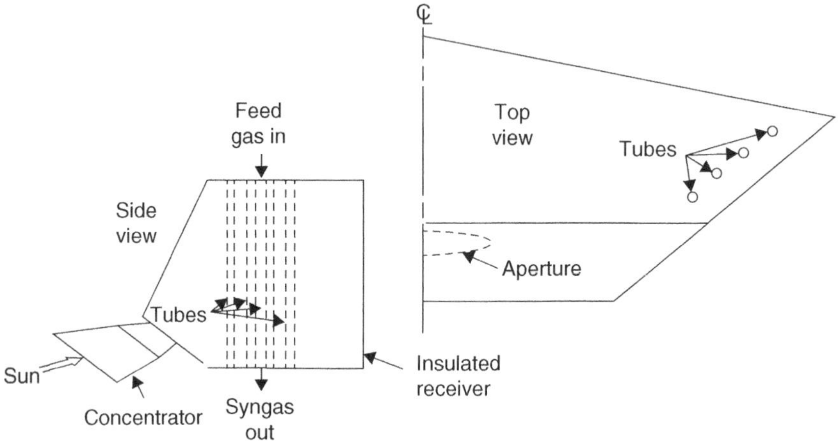

Figure 5.8 Tubular solar reformer. Source: Epstein [13]/M. Epstein.

Reforming of Methane) project, completed at the end of the 1980s, showed how a reformer functions when heated air from a solar tower power plant is used to power it. At 1,000°C and 9 bars, a maximum airflow of 0.36 kg/s was produced. It provided a 170 kW tubular reformer with a packed bed of catalyst, enabling the production of an industrial-grade synthetic gas with a chemical conversion between 68% and 93%. To ensure that the process runs continuously, the decoupled (receiver-reactor) operation can be employed as an intermediate heat storage method. In addition, a 480 kW indirect solar heating tube reactor was tested [13] atop the Weizmann Institute of Science tower in Israel. It resembles a traditional reformer in that it has a hollow in which tubes are inserted (Figure 5.8). The reaction's energy needs are met by concentrating solar energy that is absorbed through a circular aperture. Eight 4.5 m long tubes with a 6.35 cm diameter are housed in the hollow. For the manufacture of synthetic gas at 800°C using steam or dry reforming, the working pressure is 18 bars. The idea of the "on ground" reformer was taken into consideration because this type of reactor represents a sizable mass for the receiver. It employs a beam down optics optical system at the top of the tower, which theoretically might enable solar hybridization of a standard reformer for continuous operation. In fact, the system can run on burners when there is little sunlight during the day and on solar power at night. The direct heating reformer based on foamed ceramics (Figure 5.9) was one of the last direct heating reactors to be designed, produced, and tested. The basic idea is to directly heat a substrate with a large specific surface area to maximize both solar radiation absorption and catalytic processes. For a pressure of 15 bars, 950°C is the predicted temperature [15, 16].

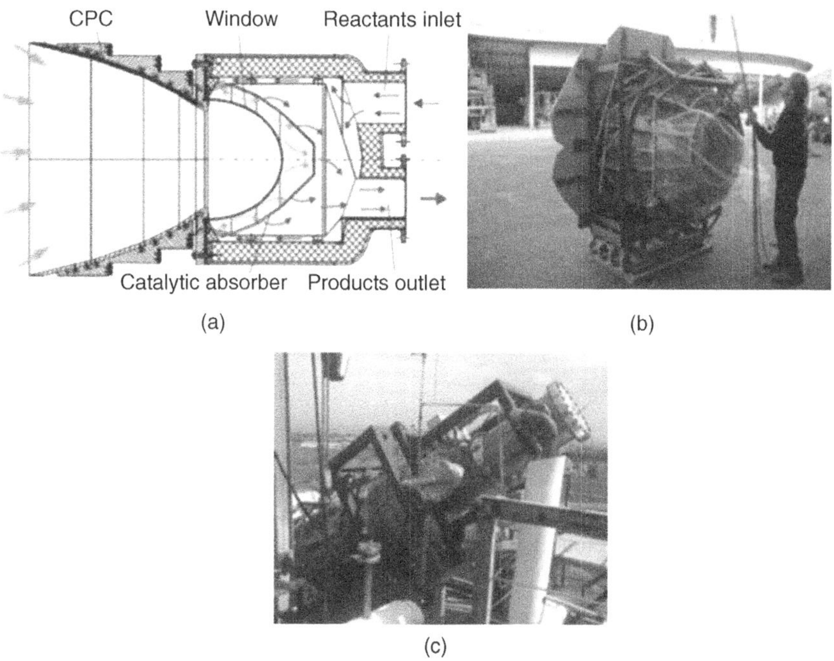

Figure 5.9 Solar reformer with direct heating: (a) schematic diagram, (b) assembled reactor, and (c) reactor installed on top of the solar tower at Weizmann Institute of Science in Israel. Source: Agrafiotis et al. [14]/with permission of Elsevier.

5.5 Economic Aspects

In solar-based cracking/reforming systems, the cost of natural gas has a clear impact on the cost of hydrogen generation. In addition, the utilization of the jointly produced carbon black affects the cost of hydrogen in the methane cracking process. It is particularly challenging to predict the price of installing solar panels because there hasn't been any industrial deployment to date, and the figures are predicted using information on commercial power plants. For instance, Figure 5.10 illustrates how the cost of hydrogen generation changes based on the price of carbon black sold for conventional reforming, solar reforming, and methane cracking. Natural gas costs 0.24 USD/Nm^3 for the study's purposes. It should be highlighted that solar cracking technologies could compete with traditional ones for the selling of carbon black at a price of roughly $1/kg. In contrast, research indicates that solar reforming could become cost-competitive at a natural gas price of approximately 0.35 USD/Nm^3. The carbon tax may in the future have an impact on these outcomes.

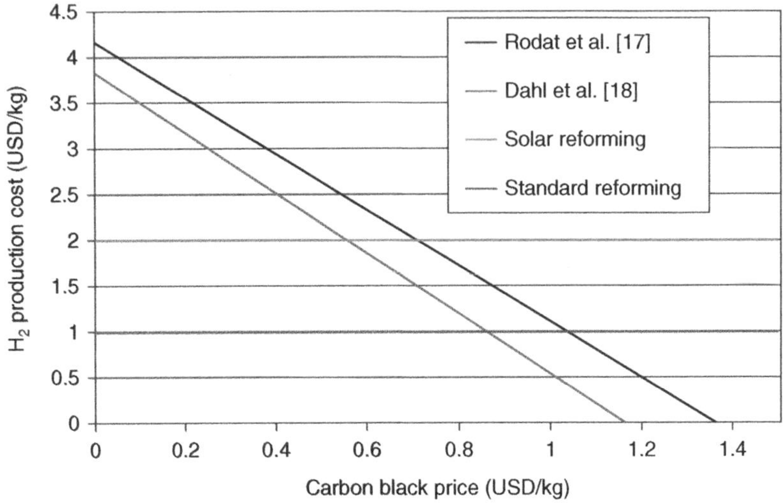

Figure 5.10 Hydrogen production cost depending on the carbon black price compared to solar reforming and conventional reforming. Source: Adapted from Rodat et al. [17].

The pyrolysis and reforming of gaseous hydrocarbons were discussed in the first section; reforming can be equated to pyrolysis followed by gasification. The conversion of solid carbonaceous resources also uses the steps of pyrolysis and later gasification.

5.6 Solar Pyrolysis and Gasification of Solid Carbonaceous Materials

The reaction pathways can be highly intricate depending on the original carbonaceous charge's makeup. However, three key phases can be distinguished (Figure 5.11). The solid is first broken down into char, tars, and noncondensable gases during the pyrolysis (or devolatilization) phase, which occurs between 300 and 1,000°C. Mineral materials can also be retrieved from ashes.

The heating rate that influences the distribution of solid and gaseous products during this stage is very important. The sun energy's high fluxes and temperatures make it easier to produce gas. Depending on the raw material used, the reaction may be slightly exothermic or endothermic. Second, an oxidizing agent (such as water vapor or CO_2) gasifies the resultant char in accordance with the principal processes [19, 20].

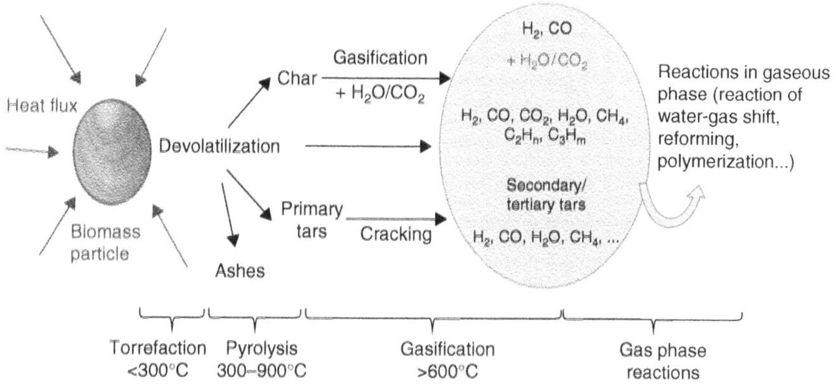

Figure 5.11 Thermochemical conversion of biomass. Source: www.iste.co.uk/flamant/energy.zip.

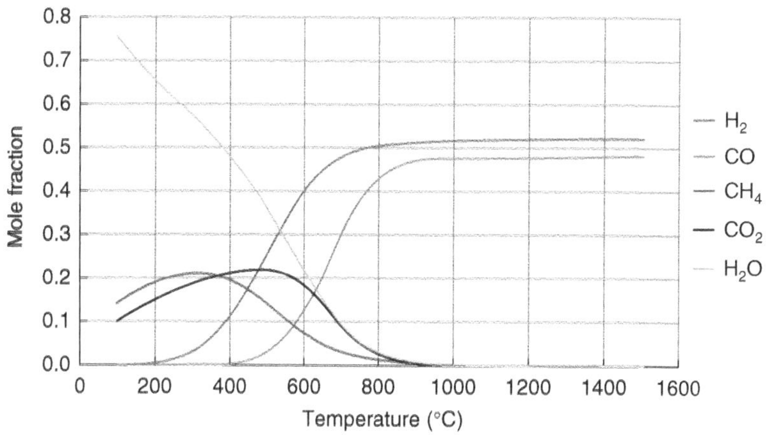

Figure 5.12 Evolution of the composition at thermodynamic equilibrium of synthetic gas during wood gasification with water depending on the temperature ($C_6H_9O_4 + 2H_2O$). Source: Adapted from Bellouard et al. [21].

As shown in Figure 5.12, the thermodynamic equilibrium of the ($C_6H_9O_4 + 2H_2O$) combination essentially results in the creation of H_2 and CO above a temperature of 1,000°C.

Figure 5.13 depicts the energy balance of the gasification reaction for a typical biomass (beech wood, formula $CH_{1.66}O_{0.69}$). Solar gasification, similar to solar reforming, is a suitable way to store solar energy in gaseous form because the resulting synthetic gas contains 26% solar energy (U = 1.26). With 47% of the

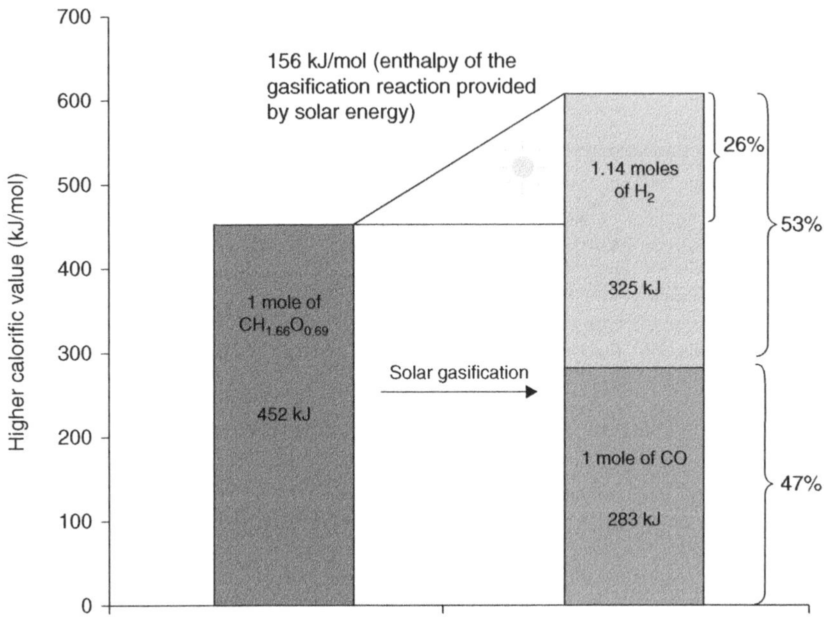

Figure 5.13 Energy balance of solar gasification of biomass (beech wood, formula $CH_{1.66}O_{0.69}$). Source: Adapted from www.iste.co.uk/flamant/energy.zip.

energy in the form of CO and 53% in the form of H_2, it has a balanced mix of CO and H_2.

5.6.1 State of the Art

The gasification of biomass using solar energy was investigated in various reactors. The original pieces predate the 1980s. In particular, packed-bed reactors, fluidized/entrained bed reactors, and spouted bed reactors are among the many types of reactors that have been constructed. A 150 kW packed-bed solar reactor was developed and tested for gasifying various carbonaceous materials, such as wood, waste, and tires, at temperatures reaching up to 1,100°C. It merely consists of two overlapping cavities: the first one is occupied by the material to be gasified, while the second one acts as a solar absorber (Figure 5.14). Due to its lengthy residence time, this type of reactor can be employed to achieve large conversion rates. In terms of both content and particle size, it can accept a wide range of raw materials. On the other hand, the packed bed's thickness causes large temperature differences in its height, which makes it harder to control reaction products and makes tar formation easier. It furthermore runs in batch (discontinuous) mode. It is possible to achieve

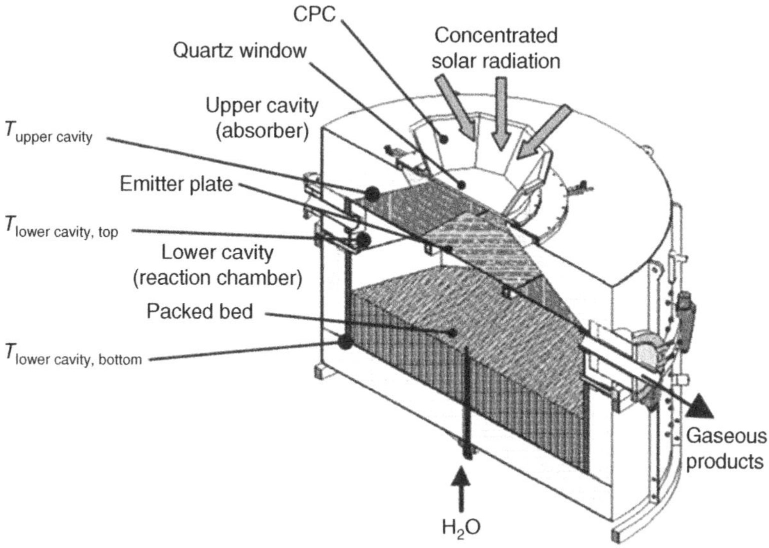

Figure 5.14 A 150 kW packed-bed reactor for solar gasification. Source: Piatkowski and Steinfeld [22]/John Wiley & Sons.

solar energy conversion efficiencies between 22% and 35% with a maximum energy upgrade factor of 1.3.

Different circulating bed concepts (fluidized beds, entrained beds, and cyclones) were developed in direct or indirect heating modes to reduce temperature gradients. The disadvantage of this kind of technology is that it often needs small particles (less than 1 mm). A spouted bed reactor is an additional alternative that is taken into account. The reactor shown in Figure 5.15 can handle various biomasses with particle sizes larger than millimeters at temperatures as high as 1,400°C. It has a vertical cone chamber with the capacity to absorb solar light directly or indirectly [23]. A hopper that transports the particles to the upper half of the cavity is used to supply biomass. At the base of the cone, where the oxidizing agent (CO_2, H_2O) is injected, the particles fall naturally due to gravity. This spout's effect is to send these particles into a cyclic motion until the reaction is finished. As a result, the solid's residence period is extended, and the stirring promotes mass and heat transfers. It is discovered that the initial energy of the biomass has an upgrade factor of 1.24. To keep the system running during the times of low sunlight, oxygen is added to water or carbon dioxide injections to enable the burning of some of the biomass [24, 25].

Research on gasification in molten salts (Li_2CO_3, Na_2CO_3, and K_2CO_3) at 1,200 K is being done in order to keep the solar reactor running even at night. To evaluate a heat storage method, an intermediate fluid can be utilized to improve heat transfers and also has catalytic capabilities. An area that needs

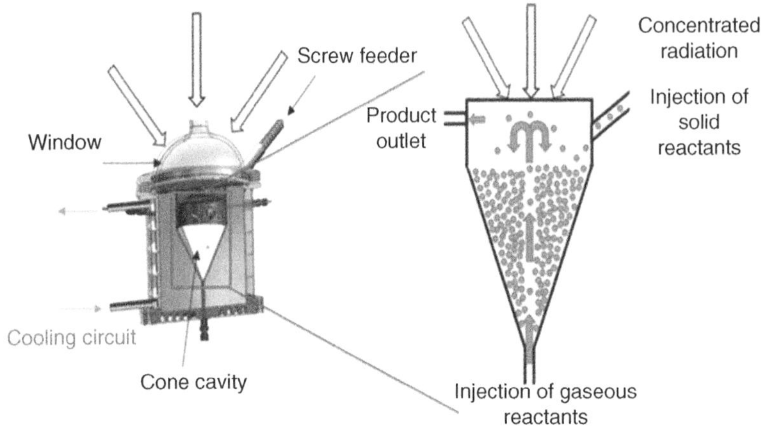

Figure 5.15 3D view and schematic representation of the spouted bed reactor (1.5 kW). Source: www.iste.co.uk/flamant/energy.zip.

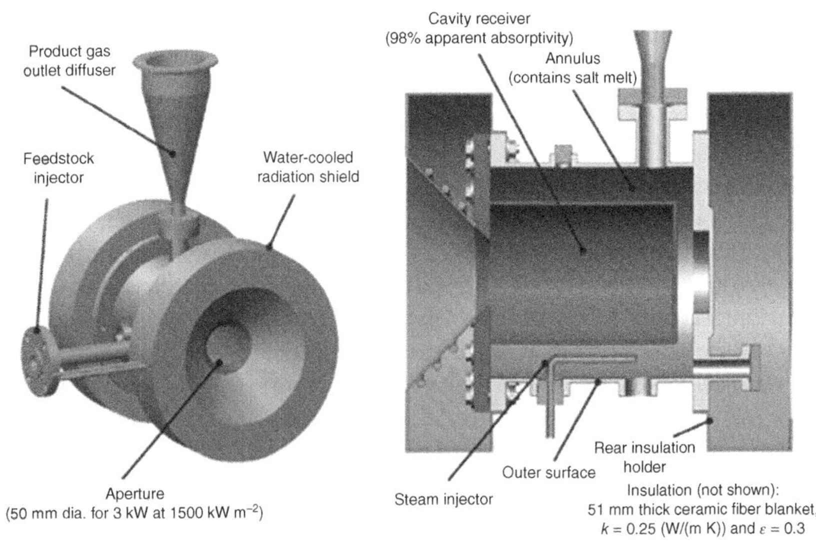

Figure 5.16 3D isometric view and sectional view of the solar reactor for the gasification of cellulose into molten salts. Source: Hathaway and Davidson [26]/American Chemical Society.

to be improved is salt stability at high temperatures. The prototype shown in Figure 5.16 is a 3 kW model. It consists of a double-walled receiving cavity where molten salt flows through while cellulose and water vapor are introduced. For a 78% carbon conversion, an energy upgrade factor of 0.9 is obtained.

5.6.2 Economic Aspects

It is unclear whether solar gasification of carbonaceous materials is more affordable than using conventional gasifiers. The conventional technique is not subject to the carbon tax since the autothermal process of biomass gasification, which uses a portion of the charge to produce the energy needed for the reaction, does not generate fossil CO_2. The cost of the portion of matter that is conventionally burned must cover the cost of the solar installation in order for the solar process to be economically viable, which is not the case in the current economic climate. In contrast, research has shown that solar gasification of coal can reduce CO_2 emissions by 26%. According to the authors' calculations, a hybrid solar method can create liquid fuel for a cost of between 5.5 and 5.8 USD/kg. When solar energy is available, this process uses it; however, when it is not, especially at night, it continues to run by burning coal in addition to or instead of solar energy. In contrast, the traditional method of coal gasification produces liquid fuel at a cost of 4.8 USD/kg. Other carbonaceous materials, particularly organic waste and other plastic materials, can be used for solar gasification. Thus, solar energy might offer a long-term answer to society's waste management issue [27].

5.7 Solar Fuel Production by Thermochemical Dissociation of Water and Carbon Dioxide

Water, carbon dioxide, and concentrated sun power can be used to produce synthetic fuels using solar thermochemistry without emitting greenhouse gases. High-temperature concentrated solar power can also be used as a source of heat. Hydrogen, synthetic gas (with H_2 and CO as the primary components), and the derived fuels (methanol, synthetic liquid fuels) are among the numerous solar fuels that are taken into account. Solar energy may be transported and stored for a long time in hydrogen, where it can be used in fuel cells or for direct combustion. A significant obstacle to the utilization of CO_2 emissions, such as those from industrial operations, is the conversion of CO_2 into synthetic fuel utilizing solar energy. In the medium future, it is anticipated that CO_2 recycling into liquid fuel and hydrogen production will grow into a separate economic sector. Three major industrial elements would be included in this sector: Harvesting of resources and energy. Despite not being at the industrial stage yet, numerous industrial applications currently have the capacity to supply the CO_2 needed for the launch of this industry. This area is increasing quickly, especially in terms of CO_2 collection and sequestration. Seawater can be desalinated to produce water. Concentrated solar energy, the last of the possible energy sources, may be able to provide the energy requirements of thermochemical processes.

5.7.1 H_2O and CO_2 Dissociation

Given the numerous restrictions placed on the development of the materials that make up the cells and the related poor efficiencies, photoelectrolysis among the investigated pathways appears to be the most challenging to carry out at a large scale. Due to its proximity to fuel cells, electrolysis is the most commercially viable technology, but more suitable and affordable materials are continuously being researched. This process still relies heavily on a low-cost electricity supply, and using a renewable energy source (solar or wind) is taken into account within the context of storing these erratic energies. The development of thermochemical cycles based on metallic oxides is also in an intermediate stage. Their focus is on a greater efficiency possibility, where heat is turned directly into chemical energy without first being converted into electrical energy [28].

5.7.2 Liquid Fuel Production

The many thermochemical routes for dissociating H_2O and CO_2 are discussed in this chapter; electrochemical, photochemical, and photobiological routes are not taken into account. The separation of H_2O and CO_2 by thermochemical cycles using metallic oxides appears to be the most effective of these processes because solar energy is used directly as heat, without losses associated with the transformation into electricity (as would be the case for electrolysis), and because it may have a higher energy efficiency than thermolysis because of a lower temperature and fewer radiative losses. The solar reactors for implementing different sorts of oxide-based cycles are given [29].

5.7.3 Direct H_2O and CO_2 Thermolysis

According to the following reaction, thermolysis is the process of breaking a bond with an oxygen atom at high temperature. Figure 5.17 shows the fluctuations in enthalpy and free energy as a function of temperature and air pressure. Also included are the temperature ranges that correspond to the various approaches under consideration.

Because CO_2 thermolysis is so endothermal, it cannot happen on its own below 3,000°C. For H_2O thermolysis, this temperature barrier is even greater. Simple data appear to support the idea that working in these extreme temperatures is irrelevant [30, 31].

- Because the materials that can withstand such temperatures are expensive and scarce, designing a reactor that can tolerate prolonged thermal cycles

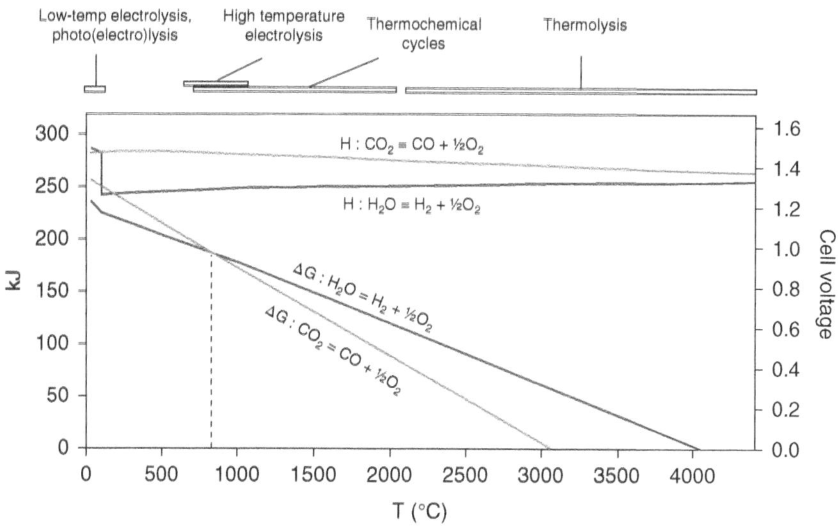

Figure 5.17 Evolution of enthalpy and free energy associated with CO_2 and H_2O thermolysis as a function of temperature. Source: Adapted from www.iste.co.uk/flamant/energy.zip.

is a costly problem. A method to separate H_2 (or CO) and O_2 (a potentially explosive mixture) at a very high temperature, or by quickly quenching this gas mixture, can be used to stop the reaction from reversing (i.e., recombination of products). No technology exists now that can address these problems.

- Expecting high energy efficiency from this kind of procedure is unreasonable because radiative losses depend on the fourth power of temperature.
- Concentrated solar power is the only renewable energy that might be used to generate such high temperatures. Although water and carbon dioxide are molecules that absorb a portion of solar energy, it is challenging to heat a gas directly with solar radiation.

To make the process more complex, one may add, for instance, particles for radiation absorption and diffusion. Despite these challenges, a patent was filed for the technology. This procedure depends on how the CO_2 absorption spectra change with temperature. Indeed, based on their observations, when the gas is preheated to 1,800 K, solar energy is properly absorbed and causes CO_2 to be photolyzed into CO, O, and O_2. By adding a large amount of CO_2 at room temperature, which combines with O to produce additional CO, the mixture is cooled. According to reports, around 6% of CO_2 is converted into CO_2. This technique pushes thermolysis to its maximum because, despite the reaction starting at a high temperature, focused photons, not the temperature itself, are what actually set it off.

5.8 Thermochemical Cycles Principle

The decomposition of this reaction into multiple steps utilizing thermochemical cycles is one way to get over the challenges that arise with direct thermolysis. This approach, aimed at producing hydrogen from water using heat generated by a nuclear reactor (up to 950°C), was initially developed in the 1960s and gained significant attention from laboratories during the first oil crisis in 1973 After that, interest waned before experiencing a significant resurgence in the 2000s, in part because of concentrated solar energy's potential to produce heat at extremely high temperatures. The cycles based on oxidation–reduction processes of metal oxides are the ones that have received the most attention for use at high temperatures (over 1,000°C). Equation [3.3] describes the first stage, which involves reducing the metal oxide (MOx) at a high temperature to produce a reduced oxide and oxygen, and equation [3.4] describes the second stage, which involves reoxidizing this product in the presence of CO_2 (or H_2O) to regenerate the metal oxide and produce CO (or H_2).

The estimated cost of hydrogen produced using this cycle ranges between 46 and 66 USD/GJ (5.5–3.9 USD/kg). While it is currently more expensive than hydrogen produced through electrolysis using nuclear electricity, it holds greater potential for cost reduction due to the sharp decline in infrastructure costs. The manufacture of syngas with a regulated H_2/CO ratio, which can be used for the synthesis of liquid fuels (synthetic fuels), is made possible by the separate or simultaneous production of H_2 and CO. This phase of the pathway's development is crucial because it secures the transition to a future economy based on fuel cells and hydrogen as well as the storage of hydrogen in the form of a fuel with a high energy density per unit weight. Indeed, CO and H_2 have high energy densities per unit of weight but low densities per unit of volume (Figure 5.18; notice that diesel is similar to gasoline and DMF, or third-generation biofuel; diesel is also near to DMF).

The Fischer–Tropsch catalytic synthesis, which has been commercially used since the 1920s for the generation of liquid fuel from syngas obtained by the gasification of carbon or hydrogen produced by steam reforming of natural gas, is the most "direct" way to obtain liquid fuels once the syngas mixture has been obtained. Temperature, pressure, the makeup of the syngas used, and the catalyst all affect how long the carbon chain is, which in turn affects the fuel quality. Additionally, the independent creation of the two syngas constituents makes it easier to fine-tune its composition and affects the selectivity to one of the products of the

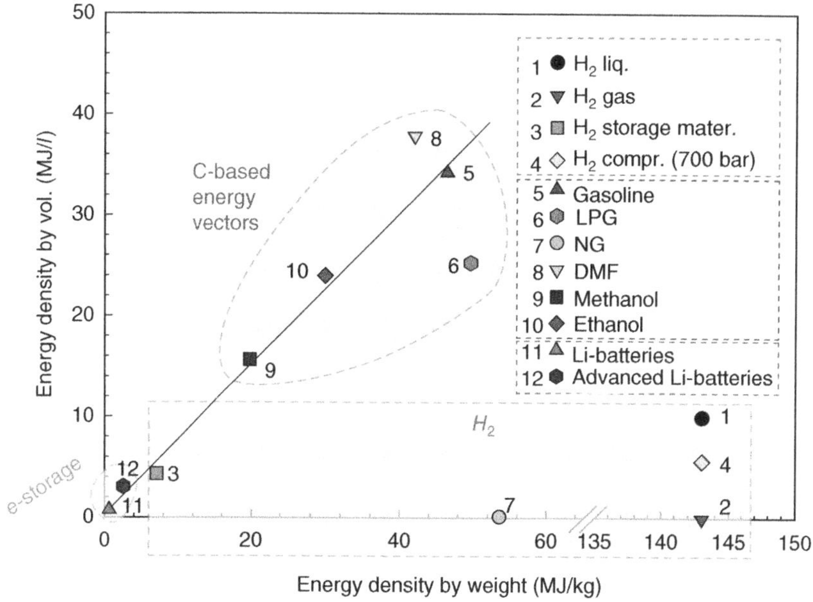

Figure 5.18 Comparison of energy densities by weight and by volume for various energy vectors. Source: Centi and Perathoner [32]/John Wiley & Sons.

Fischer–Tropsch process. Finally, sulfur poisoning of catalysts is avoided. Different cycle types can be distinguished, particularly in terms of the technology that will be used to use them. Cycles with more than two phases have the advantage of requiring lower temperatures, which are compatible with a nuclear heat source, but they also include more difficult material flow, separation, and corrosion management issues, which are not covered in this article. Regarding the two-step cycles, a distinction can be drawn between cycles where the reduction step results in either a new species (stoichiometric cycle) or a partial reduction of the oxide (nonstoichiometric cycles). Then, one can obtain this species either gaseously (cycles with volatile oxides) or not (cycles with nonvolatile oxides). The productivity of nonstoichiometric cycles in H_2/CO per mass of oxide is lower. The cycles that use volatile oxides produce a gaseous product that readily recombines (with O_2) and needs to be recovered by condensation and filtration, but whose specific surface area is substantial (many dozens of m_2/g) and is renewed at each cycle. In addition to these technological requirements, the thermodynamic characteristics of these cycles are crucial since they will be used in a theoretical exergy analysis to establish the maximal efficiencies. The mass and energy diagrams (Figure 5.19) show the different sources and wells.

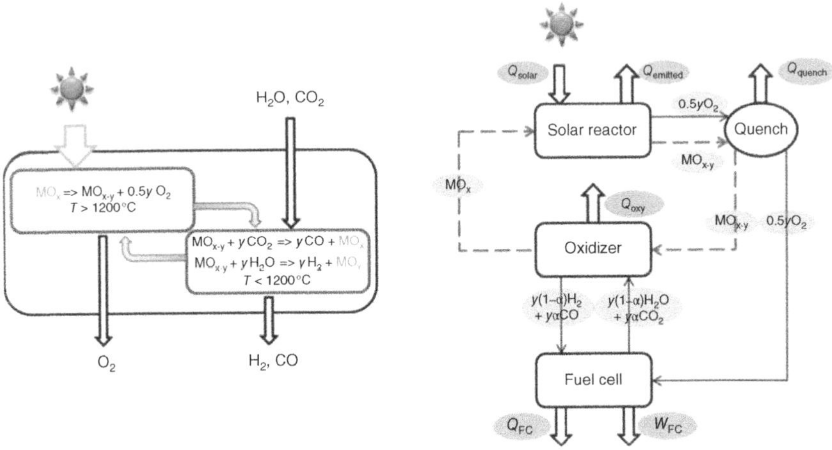

Figure 5.19 Mass and energy diagram for a thermochemical cycle. Source: www.iste.co.uk/flamant/energy.zip.

5.9 Cycles with Volatile Oxides

Depending on the heat recovery level, between 35% and 50%. With a temperature range of 1,800–2,100 K and a zero-order reaction, the kinetics of the reduction of zinc oxide was investigated using a solar-driven thermogravimetry system. Other options were taken into account, particularly the aerosol reduction, which increases the pre-exponential component by several orders of magnitude. The direct irradiation of the reactant is preferable, and the separation of Zn(g) and O_2 to avoid recombination are key factors in this stage's overall efficiency. Although the energy thus collected from the products cannot be reused in the process, a quick quenching is often still considered the best option (zinc vaporization temperature is approximately 1,200 K and its recrystallization follows a heterogeneous process from nucleation sites). Zn reoxidation occurs in two stages, the first of which is quick and corresponds to the oxidation of the available zinc surface, and the second of which is slower and is slowed down by diffusion occurring inside zinc particles (passivation of particle surface). The response rate is directly influenced by particle size, with smaller particles restricting the diffusion process. Contrary to temperature, the concentration of the oxidizer has no effect on the kinetics of oxidation involving simply H_2O or CO_2. In terms of kinetics, the reactions involving H_2O and CO_2 are fundamentally different from one another: for comparable particles, the activation energies of the second step were estimated to be 83.77 kJ/mol and 16,225.3 kJ/mol, respectively. The co-reduction of H_2O and CO_2 was researched despite this significant discrepancy. It was found that the composition of the input gas had a significant impact on the syngas's quality. Studying the oxidation of gaseous Zn by CO_2 or H_2O revealed 10 times greater kinetics than for solid Zn. Despite this speed, it is unlikely that this strategy will be applied generally because it is challenging to apply. Zn must be condensed even though it is initially created in the gaseous state in order to prevent recombination and allow it to be separated from O_2. The energy released during this condensation is then wasted through quenching. The following are the key traits of this cycle as determined by earlier studies: the temperature of the reduction stage is high (over 2,000 K);

- The reoxidation step is constrained by a passivation phenomenon brought on by the creation of a zinc oxide layer;
- The quenching of gases from >2,000 to >1,200 K is a crucial stage: its quality is critical for the following step (size of zinc particles and purity) as well as energy efficiency of the process.

The temperature of the reduction stage can be lowered, thanks to the addition of carbon to zinc oxide. In fact, carbon is a reducer that speeds up the reaction and causes CO to be produced in the initial phase .The best reducers are charcoal and activated carbon, and the type of coal is crucial for the reduction. Two solid-gas reactions make up the retained reaction mechanism.

SnO_2/SnO. Tin may oxidize to various degrees, +II or +IV; therefore, a number of redox pairs, including SnO_2/SnO, SnO_2/Sn, and SnO/Sn, can be taken into consideration. Souriau patented a first cycle with three phases in 1971 with the intention of utilizing nuclear reactor heat as a source of energy. SnO_2 is reduced at a high temperature to SnO, which disproportionates at 600°C to form liquid metallic tin and SnO_2. The oxide is then renewed when liquid tin combines with water to create hydrogen.

In a study by Charvin, dismutation was seen, but the development of two phases was not, impeding the recovery of Sn (l), rendering the remainder of the procedure ineffective. Since then, the SnO_2/SnO cycle has received attention; the first step is similar to the Souriau cycle; however, the second phase skips the dismutation stage and employs SnO directly as a reducer.

It has been established that SnO is not always more effective than metallic tin for the reduction of water. PROMES (WO 2008113944) filed a patent for the use of this cycle in hydrogen production. Product recombination limits the reduction step in a manner similar to zinc. Le Chatelier's principle states that lowering the pressure in the reactor chamber encourages the generation of gaseous species and permits lowering the reaction's needed temperature (1,900 K at 20 kPa as opposed to 2,342 K at atmospheric pressure). The partial pressure of O_2 can be reduced thanks to the dilution by a neutral gas, which in turn restricts recombination. In a similar manner, the reaction's necessary temperature is likewise lowered. The condensation of SnO and the lowering of temperature below the reaction temperature are finally made possible by a quick quenching of the gases. From the nucleation sites, condensation proceeds through a heterogeneous process. Numerous considerations should be made throughout the reoxidation procedure. The working conditions (SnO or Sn/SnO_2) may affect the reduced oxide powder's composition. In fact, during the temperature rise or the first cooling, partial SnO dismutation may take place. Similar to how the reoxidation reaction is a solid/gas reaction, particle morphology dominates. From this perspective, recovering SnO through condensation enables the production of powders with grains that are several dozen nanometers or smaller, producing a large amount of specific surface area. Finally, not all kinetics are susceptible to CO_2 or H_2O reduction. The activation energy for the reduction of H_2O from 798 K is 51 ± 7 kJ/mol, whereas the activation energy for the reduction of CO_2 from 973 K is 88 ± 7 kJ/mol. According to Abanades, the zinc cycle has a number of benefits.

The temperature of reduction is lower than the temperature of the zinc cycle and might be further lowered by the drop in chamber pressure. Lower operating temperatures directly imply greater energy efficiency and, thus, better process economic viability. A 72% pure product is achieved, compared to 48% for ZnO; The oxide dissociation rate is significant and is less dependent on the gas quenching at the reactor exit than for zinc. In fact, whereas SnO has a boiling point of 1,800 K, zinc has a boiling temperature of 1,180 K, making it easier for SnO to condense. SnO is a useful water reducer, to sum up.

5.10 Nonvolatile Oxide Cycles

The benefit of nonvolatile oxide cycles is that during the reduction step, no gaseous species are formed. This eliminates the need to manipulate the substance in between cycle steps, and if melting doesn't take place, the two reactions (hydrolysis and reduction) can happen simultaneously in the same reactor with only one temperature change. The cycle based on iron oxides (magnetite/wüstite) was the first nonvolatile oxide cycle to be proposed. At 2,100 K, the magnetite (Fe_3O_4) is decreased, and at 1,811 K, it melts into a liquid state. At a minimum temperature of 450 K, the hydrolysis process takes place. According to experimental studies , the hydrolysis stage has strong reactivity but is constrained by the development of a passivating oxide layer. This entire reduction is demonstrated under inert gas.

The studies also showed that quenching was required after magnetite reduction to stop wüstite from dismutating into iron and magnetite at temperatures below 570°C. Based on the HHV of H_2, the predicted energy efficiency of this cycle is 35%:

Mixed compounds were created based on the formation of an iron oxide solid solution with an M_3O_4/MO redox pair that had improved reducibility at lower temperatures and improved thermal stability of the material while it was still in the solid phase in order to lower the temperature of the cycle. The $MxFe_3$-$xO_4/MxFe_1$-xO redox pair so produced can be reduced at a lower temperature and endures the high-temperature step while still in the solid state. A ferrite is created when the dopant is incorporated into the magnetite's spinel structure. The doping causes a structural deformation of the spinel structure, which promotes oxygen movement, further facilitating the decrease. Numerous dopants, including Ni, Mn, Zn, Co, Cu, and Mg, were tried. The CeO_2/Ce_2O_3 system is used in a different cycle. The heat reduction of cerium +IV into cerium +III and subsequent hydrolysis are the two processes that make up the cycle.

At 1,950°C and reduced pressure, a complete reduction can be achieved in the liquid phase. Ce_2O_3 has excellent reactivity (100% conversion) and a quick hydrolysis reaction from 400°C (complete reoxidation in five minutes as opposed to 30% in 120 minutes for FeO), and the size of the particle has little to no impact because of oxygen diffusion in the crystal. Even still, the temperature of reduction is still very high, and CeO_2 was only partially sublimated. This cycle's energy efficiency is 40%. To increase the rate of ceria reduction and carry out this phase at lower temperatures, ceria doping with another metal might be taken into consideration.

5.11 Nonstoichiometric Oxide Cycles

5.11.1 Ferrite-Based Cycles

The early studies on the topic centered on ferrites, and only these have progressed to the point of a 100 kW demonstrator. As such, using the Fe_3O_4/FeO cycle is challenging. On the other hand, the reaction takes place in the liquid phase even though the temperature needed for Fe_3O_4 reduction is extremely high and above its melting point. After that, the product will need to be ground to have a particular surface area. Additionally, the resultant FeO must be quenched because it dismutes into Fe_3O_4 and Fe above 570°C. It is possible to get around these restrictions by replacing some of the iron atoms in the original crystal with different metals. The goal is to introduce a metal that is more easily reducible but less reactive to water (in the case of cobalt). At lower temperatures, the resulting doped ferrites (of the $(Fe1\text{-}xMx)_3O_4$ type, where M = Mn, Co, Mg, Ni, Zn, etc.), can partially undergo reduction to the $(Fe1\text{-}xMx)_3O_4$- form while maintaining their water reactivity. Nickel ferrites seem to perform well in terms of hydrogen production and thermal stability for a reduction temperature below 1,500°C after the many investigations done on the issue. However, because sintering is unavoidable, no ferrite possesses adequate cycle capacity. The coating of a ceramic substrate was extensively researched to prevent this problem. The findings show that monoclinic zirconia (m-ZrO_2) and yttrium-stabilized zirconia exhibit the best performances, with reduction rates below 30%. The hydrogen mass productivity of this kind of solution is low since the substrate may make up 90% of the weight of the reactant material.

5.11.2 Ceria-Based Cycles

In the automotive sector, cerium oxide (ceria, CeO_2) is frequently employed as a catalyst for the oxidation of exhaust gases. This application is dependent on the crystal's substantial oxygen ion diffusion, which may also be helpful for the hydrolysis stage. In 2006, Abandes and Flamant suggested the first cycle based on ceria. The CeO_2/CeO_2 system is involved in the partial reduction of ceria that results from reducing the temperature of reduction (or nonstoichiometric reduction). The reduction capacity of ceria can be improved by doping it with a trivalent or tetravalent cationic element. Numerous studies on the formulation, production, and use of doped ceria have demonstrated an interest in zirconia doping, for instance. Comparatively, the examined nickel-doped ferrites provided improved cycles and kinetics of hydrogen synthesis, with a maximum output of 250 mol/g oxide as opposed to 333 mol/g oxide for the ceria-zirconia. Both the material's structure and stability at high temperatures are crucial. These two characteristics could be a product of the powder synthesis technique or the structuring done a posteriori. The usage of reticulated foams with dual-scale porosity as porous structures made of ceria has lately been the subject of an interesting concept evaluation. The entire reactant can be heated effectively thanks to macroscopic porosity, which is itself microstructured to improve the specific surface area. It has also been investigated how biomimetics can be used to develop ordered macroporous structures. Despite recent developments, a few things still need to be worked on in order to fully utilize the benefits of ceria-based cycles:

- The formulation of ceria can be further enhanced (by using composite materials, adding more dopants, etc.).
- Synthesis techniques that provide high-temperature resistance are required in the event that the powder is used in this way. If the reactant is structured, it is necessary to develop resistant substrates, effective coating techniques, or architectured porous structure elaboration techniques.

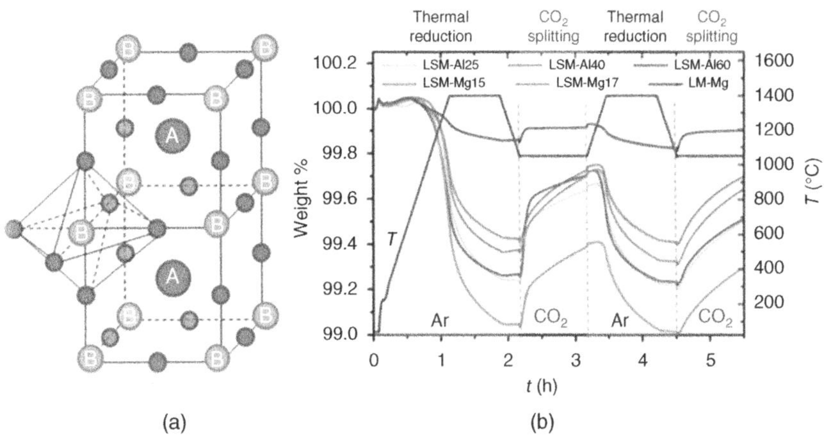

Figure 5.20 (a) ABO_3 perovskite structure; (b) reactivity of LSM perovskites doped with Al or Mg at B site during two successive cycles (reduction under Ar at 1,400°C and oxidation under CO_2 at 1,050°C). Source: www.iste.co.uk/flamant/energy.zip.

5.11.3 Perovskite Structure-Based Cycles

The usual designation for the perovskite structure is ABO_3, where A denotes a cation with a larger size than B. A and B are also linked to 12 and 6 oxygen atoms, respectively, in Figure 5.20a. Its key benefit is the capacity to change the stability of a compound by replacing specific A and B sites with different atoms.

It is feasible to achieve structural and electronic crystal deformations that may aid the displacement and escape of oxygen by carefully selecting the atoms that fill the major sites and the doping cations. This property, along with the structure's electronic stability, allows for the consideration of a large oxygen nonstoichiometry. It is therefore possible to investigate materials with tunable reduction and oxygen diffusion capacities that are greater than those of ferrites and ceria LaxSr1-xMnO_3 (LSM), a perovskite formulation, has been explored (generation of H_2 or CO over 200 mol/g, with a , temperature of reduction of 1,400°C, Figure 5.20b), and it has demonstrated a larger hydrogen production potential than CeO. These outcomes are nonetheless equivalent to those achieved for doped ceria, but they demonstrate significant room for advancement.

5.12 Solar Reactor Concepts for Cycle Implementation

Reactors that enable the two parts of the cycle to proceed in a single chamber and those whose reduction and reoxidation functions are separated can be divided into two primary groups. For the reactor to run at its most efficient, solar heat integration is crucial.

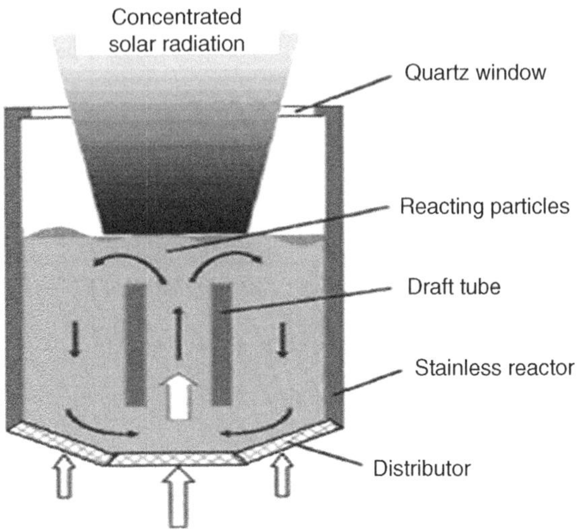

Figure 5.21 Diagram of a fluidized bed with internal circulation.

Single-chamber reactors: The bulk of approaches for implementing nonvolatile oxide cycles are grouped together in this type of reactor. The reactant material can be either separated (pure substance or coated as thin films) or structured (by covering a substrate or by shaping). To promote phenomena at interfaces, a large specific surface area is desired in both scenarios. The fluidized particle beds with internal circulation are one of the different reactor concepts (Figure 5.21). The reduction/reoxidation reactions are smooth, the potential for sintering is constrained, and the temperature of the particles is homogenized as a result of their motion. The gas flow in the cylindrical reactor's center enables particle fluidization and pushes the particles upward into the bed. When the upper portion of the bed is at the focal point of the concentration system, which enables the reduction, the gas flow rate is adjusted to the size and density of the particles (determining the conditions required for their fluidization). A portion of the heat is transferred from the top to the lower half of the bed when the particles return to the periphery at the bottom of the bed after reaching the irradiation zone. In comparison to conventional beds, a better vertical homogeneity of temperature can be achieved, and the core part requires less neutral gas flow. Irradiation ceases after reduction is finished, and some of the neutral gas flow is replaced by water vapor. For instance, this kind of reactor has been used with ZrO_2 particles coated in nickel ferrite ($NiFe_2O_4$). With a 100 kW pilot project, another monolithic reactor idea that focuses on solar-based hydrogen production has been created. The reactant material (doped ferrite) is placed atop a ceramic honeycomb structure that serves as both a hydrolysis reactor and a volumetric radiation absorber (Figure 5.22). Two identical reactors are

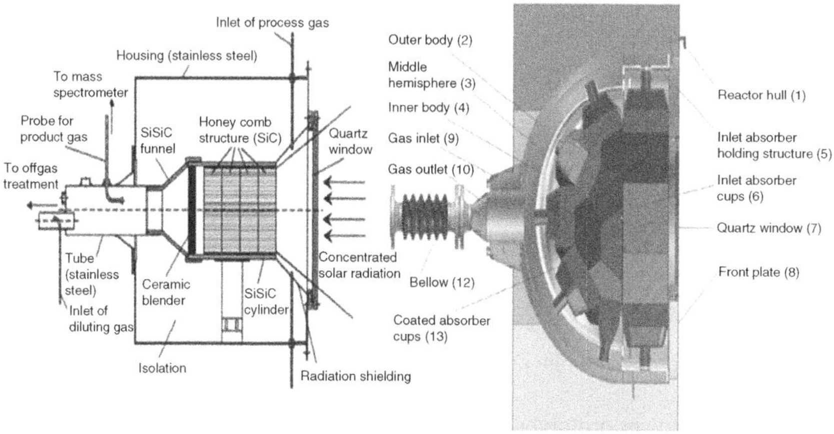

Figure 5.22 Diagrams of monolithic reactors integrating a porous structure.

placed in parallel to ensure continuous operation. Both of them host ferrite hydrolysis and reoxidation when one of them is prone to concentrated solar flux, permitting reduction, and vice versa. As a result, the reactors are fixed, with heliostats controlling the stages' alternation. More recently, other monolithic reactors with enormous porous ceria structures (reticulated foams) have been created.

5.13 Decoupled Reactors

This category includes all reactors that create H_2/CO in a second reactor at a lower temperature while using a solar receiver for the reduction stage. As a result of the cavity's being precisely tuned for this reaction, the heating/cooling cycle is eliminated, and the assembly might theoretically run continuously. The reactors used to reduce volatile oxides are the most typical example, Introduced the idea of a revolving cavity reactor (Figure 5.23). The ZnO particles, which are fed into the cylindrical cavity by a screw feeder in the axial part, are dispersed across the walls as a result of rotation, preventing the creation of hot spots because of the nonuniformity of the incoming flux. To prevent particle buildup on the quartz window's rear face, a steady flow of gas is injected there. The resultant gases are quenched (cooled) by bulk injection of argon at room temperature through an annular portion around the screw feeder. In the end, filtering is used to recover the zinc particles (50% purity). Another revolving reactor design was created, allowing for continuous operation while injecting ZnO particles and recovering Zn products as a fine powder at the exit.

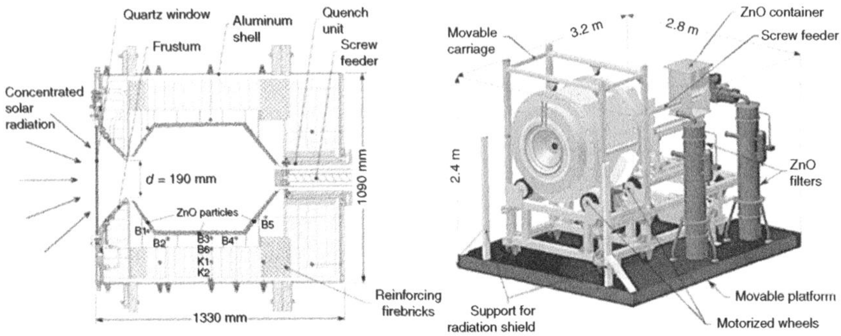

Figure 5.23 100 kW reactor for ZnO dissociation.

The highest power prototype for the reduction of metal oxide was built during the Solzinc European project (Figure 5.24). This reactor implements the carboreduction of zinc oxide, thus reducing notably the thermal constraints and the technological difficulties of the gas quench to the detriment of carbon balance. The reactor is composed of two cavities. The upper part of the reactor, composed of SiC plates, receives and absorbs radiation and reemits it to the lower part containing a packed bed of particles, heated by the radiation of the separation plate. This allows a significant production of reduced particles without risking

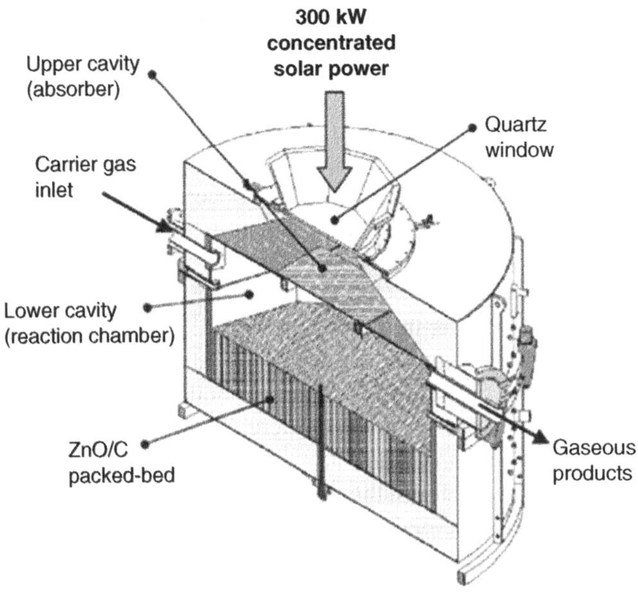

Figure 5.24 300 kW reactor for the carboreduction of ZnO.

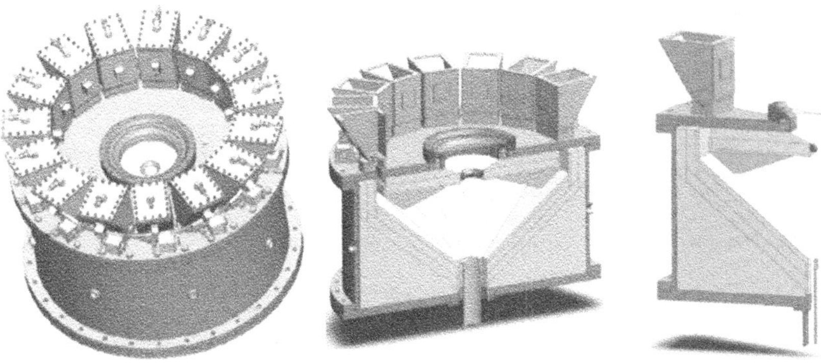

Figure 5.25 Diagram of the reactor with gravity-driven injection of particles. Source: www.iste.co.uk/flamant/energy.zip.

their deposition on the quartz window of the reactor. A 50 kg/h production of 95% pure zinc was obtained, for an operating temperature of 1,300–1,500 K and a 30% efficiency of the process.

For the thermal dissociation of ZnO, a new reactor design with a vertical irradiation axis (gravity-fed solar thermochemical reactor; Figure 5.25) was proposed. Particles are fed through 15 distributors positioned around the reactor's perimeter to uniformly distribute them across the space. The refractory material plates that make up the cavity, whose slope and surface were chosen for this reason, must be covered by the particles by gravity in the shape of a curtain. During their development, the particles must be diminished and evaporated, with the gases being propelled toward the reactor's exit by a rotating flow of neutral gas. The annular area between this gas exit and the lower half of the reactor is where the particles that did not react are retrieved. Additionally, the PROMES laboratory has created technical approaches that are suited to the ongoing reduction of volatile oxides. A preliminary setup that included a spinning cavity was put together. The capability of having reactions under lower pressure in order to boost kinetics was one of the system's goals and its novelty. A second, distinct prototype that operates at decreased pressure and has a vertical irradiation axis was created. The lower portion of the insulated cavity constantly feeds compressed reactant (pellets), which is simultaneously heated by direct concentrated radiation and radiation from the cavity walls (Figure 5.26). This reactor's key benefit is a shorter distance that reduces recombination between the reduction zone and the evacuation connected to the particle filter system. For an operating pressure of roughly 20 kPa and an estimated efficiency of solar energy into chemical energy conversion of 2–3%, the powders recovered at the outlet comprise roughly 40% decreased species for ZnO and 70% for SnO_2.

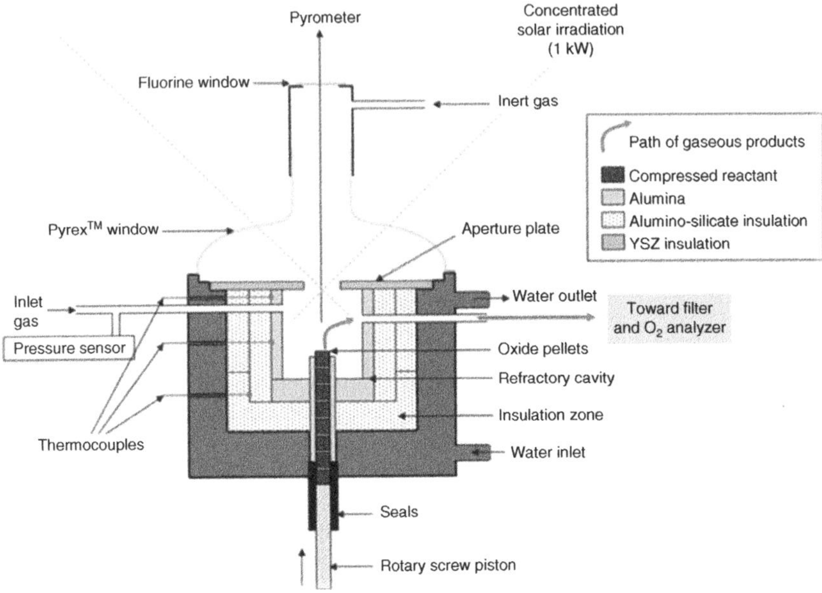

Figure 5.26 Reactor for the dissociation of volatile oxides (ZnO, SnO_2) with continuous injection of reactant in compressed form. Source: www.iste.co.uk/flamant/energy.zip.

Two solid oxide reactors with an integrated displacement of the reactant from one chamber to another were created with the goal of integrating heat between the two phases and of continuous operation. The CR5 idea (Counter-revolving-Ring Receiver/Reactor/Recuperator, Figure 5.27a) uses two chambers that are divided by a stack of revolving rings. Rotation moves the metal-oxide-coated ring surface from the reduction zone to the oxidation zone. The heated part that exits the reduction zone of one neighbor reheats the colder sections that enter the two neighbors' reduction zones, allowing a 30% heat recovery. The difficulty of this idea lies in this ring assembly, which also needs to divide the two chambers from one another. There aren't enough data available to determine the concept's actual efficacy, despite the fact that a prototype employing cobalt ferrites was used to experimentally validate it.

The particles are directed by a stationary screw in the opposite concept (Figure 5.27b), toward the upper region of the reactor, where they are exposed to concentrated radiation. Due to gravity, they fall back down into the reoxidation chamber through the hollow axis of the screw, which serves as a heat exchanger between upward and downward particles. The ability to operate constantly is highlighted by the authors in contrast to the fluidized bed, which only permits batch operation. However, this theory has not been tested, and reaching tightness

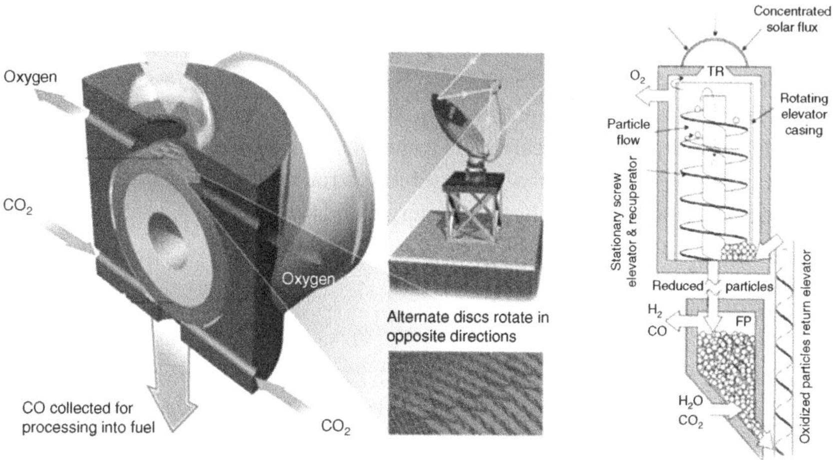

(a) (b)

Figure 5.27 (a) Diagram of CR5 reactor and (b) prototype of circulating dense bed reactor. Source: www.iste.co.uk/flamant/energy.zip.

because of stacked particles appears challenging, much as moving particles to a zone of high temperature.

5.14 Conclusion

The conversion of carbonaceous materials into gaseous or liquid solar fuels using concentrated solar energy presents a sustainable alternative to conventional fossil fuel-based processes. By utilizing solar energy for endothermal reactions such as hydrocarbon cracking, reforming, pyrolysis, and gasification, this technology offers numerous environmental and economic benefits. The primary advantage is the avoidance of fossil fuel combustion to generate process heat, which significantly reduces hydrocarbon consumption and minimizes CO_2 emissions. This, in turn, helps mitigate climate change by lowering the carbon footprint of fuel production.

Moreover, the conservation of raw materials—since they are not burned—ensures more efficient use of resources, contributing to the sustainability of energy systems. The ability to store solar energy in the form of clean, usable fuels further enhances the practicality of these systems, enabling the production of fuels with lower environmental impact. The high temperatures and radiative fluxes achieved through solar concentration allow for efficient and novel reaction pathways, making the process highly versatile and capable of producing a range of cleaner fuels.

In conclusion, the integration of concentrated solar energy into the conversion of carbonaceous materials holds great promise for creating a cleaner, more sustainable energy future. It reduces the reliance on traditional fossil fuels, curbs CO_2 emissions, and offers a pathway for storing solar energy in a useful, transportable form. Continued research and technological advancements will be critical to scaling these processes for widespread industrial applications, ultimately contributing to a more sustainable global energy landscape.

References

1 Zhang, J., Wang, H., Dai, F., and Kong, H. (2024). Thermodynamic analysis of the coal-driven solar thermochemical cycle for hydrogen production. *Applied Energy* 375: 124059.

2 Li, C., Wang, M., Li, N. et al. (2024). Solar oil refinery: solar-driven hybrid chemical cracking of residual oil towards efficiently upgrading fuel and abundantly generating hydrogen. *Energy Conversion and Management* 300: 117900.

3 Hirsch, D. (2001). The solar thermal decarbonization of natural gas. *International Journal of Hydrogen Energy* 26 (10): 1023–1033.

4 Rodat, S., Abanades, S., Sans, J.-L., and Flamant, G. (2010). A pilot-scale solar reactor for the production of hydrogen and carbon black from methane splitting. *International Journal of Hydrogen Energy* 35 (15): 7748–7758.

5 Guo, Y., Chen, J., Song, H. et al. (2024). A review of solar thermochemical cycles for fuel production. *Applied Energy* 357: 122499.

6 Xin, Y., Zhang, W., Chen, F. et al. (2024). Integrating solar-driven biomass gasification and PV-electrolysis for sustainable fuel production: thermodynamic performance, economic assessment, and CO_2 emission analysis. *Chemical Engineering Journal* 497: 153941.

7 Hirsch, D. (2004). Solar hydrogen production by thermal decomposition of natural gas using a vortex-flow reactor. *International Journal of Hydrogen Energy* 29 (1): 47–55.

8 Zarei-Jelyani, F., Zarei-Jelyani, M., Salahi, F., and Rahimpour, M.R. (2024). Application of solar energy in hydrogen production. In: *Hydrogen Production from Renewable Resources and Wastes* (ed. M.R. Rahimpour, M.A. Makarem, and P. Kiani), 281–294. CRC Press.

9 Abanades, S., Kimura, H., and Otsuka, H. (2015). A drop-tube particle-entrained flow solar reactor applied to thermal methane splitting for hydrogen production. *Fuel* 153: 56–66.

10 Christopher, S.D., Devarajan, Y., Raja, T., and Jayabal, R. (2024). A comprehensive review of biomass pyrolysis for hydrogen production in India. *Process Safety and Environmental Protection* 190: 646–662.

11 Seo, Y.-S., Shirley, A., and Kolaczkowski, S.T. (2002). Evaluation of thermodynamically favourable operating conditions for production of hydrogen in three different reforming technologies. *Journal of Power Sources* 108 (1–2): 213–225.

12 Awogbemi, O., Ojo, A.A., and Adeleye, S.A. (2024). Advanced thermochemical conversion approaches for green hydrogen production from crop residues. *Journal of Renewable Materials* 12 (1): 1–28.

13 Epstein, M. (2011). *Solar Thermal Reforming of Methane*. Zürich: SFERA Winter School.

14 Agrafiotis, C., von Storch, H., Roeb, M., and Sattler, C. (2014). Solar thermal reforming of methane feedstocks for hydrogen and syngas production: a review. *Renewable and Sustainable Energy Reviews*, 29: 656–682.

15 Taqvi, S.A.A., Kazmi, B., Naqvi, S.R. et al. (2024). State-of-the-art review of biomass gasification: raw to energy generation. *ChemBioEng Reviews* 11: e202400003.

16 Xie, Y., Wang, J., Cai, X. et al. (2016). Pressure history in the explosion of moist syngas/air mixtures. *Fuel* 185: 18–25.

17 Rodat, S., Abanades, S., and Flamant, G. (2011). Methane decarbonization in indirect heating solar reactors of 20 and 50 kW for a CO_2-free production of hydrogen and carbon black. *Journal of Solar Energy Engineering* 133 (3): 031001.

18 Dahl, J.K., Buechler, K.J., Weimer, A. W. et.al. (2004). Solar-thermal dissociation of methane in a fluid-wall aerosol flow reactor. *International Journal of Hydrogen Energy* 29(7): 725–736.

19 Nzihou, A., Flamant, G., and Stanmore, B. (2012). Synthetic fuels from biomass using concentrated solar energy–a review. *Energy* 42 (1): 121–131.

20 Singh, S.K. and Tiwari, A.K. (2024). Solar-powered hydrogen production: advancements, challenges, and the path to net-zero emissions. *International Journal of Hydrogen Energy* 84: 549–579.

21 Bellouard, Q., Rodat, S., Grateau, M., and Abanades, S. (2020). Solar biomass gasification combined with iron oxide reduction for syngas production and green iron metallurgy. *Frontiers in Energy Research* 8: 66.

22 Piatkowski, N. and Steinfeld, A. (2011). Solar gasification of carbonaceous waste feedstocks in a packed-bed reactor – dynamic modeling and experimental validation. *AIChE Journal* 57 (12): 3522–3533.

23 Sun, J., Bai, B., Yu, X. et al. (2024). Thermodynamic analysis of a solar-assisted supercritical water gasification system for poly-generation of hydrogen-heat-power production from waste plastics. *Energy* 307: 132606.

24 Nawaz, A., Haddad, H., Shah, M.A. et al. (2024). Fueling sustainability: co-pyrolysis of microalgae biomass and waste plastics for renewable energy and waste mitigation. *Biomass and Bioenergy* 187: 107303.

25 Hakami, O. (2022). Thermocatalytic and solar thermochemical carbon dioxide utilization to solar fuels and chemicals: a review. *International Journal of Energy Research* 46 (14): 19929–19960.

26 Hathaway, B.J. and Davidson, J.H. (2020). Solar steam gasification of cellulose in a molten alkali salt: salt chemistry and reactor performance. *Energy & Fuels* 34 (2): 1811–1821.

27 Zueva, S., Ferella, F., Corradini, V., and Vegliò, F. (2024). Review of organic waste-to-energy (OWtE) technologies as a part of a sustainable circular economy. *Energies* 17 (15): 3797.

28 Xu, D., Gu, X., and Dai, Y. (2023). Concentrating solar assisted biomass-to-fuel conversion through gasification: a review. *Frontiers in Energy Research* 10: 1029477.

29 Saravanakumar, A., Vijayakumar, P., Hoang, A.T. et al. (2023). Thermochemical conversion of large-size woody biomass for carbon neutrality: principles, applications, and issues. *Bioresource Technology* 370: 128562.

30 Wang, J., Al-attab, K.A., and Heng, T.Y. (2024). Optimization of solid oxide fuel cell system integrated with biomass gasification, solar-assisted carbon capture and methane production. *Journal of Cleaner Production* 449: 141712.

31 Abanades, S., Rodat, S., and Boujjat, H. (2021). Solar thermochemical green fuels production: a review of biomass pyro-gasification, solar reactor concepts and modelling methods. *Energies* 14 (5): 1494.

32 Centi, G. and Perathoner, S. (2011) CO2-based energy vectors for the storage of solar energy. *Greenhouse Gases: Science and Technology* 1 (1): 21–35.

6

Concentrating Photovoltaic (CPV) Systems and Applications

6.1 Introduction

Using an optical assembly, concentrating photovoltaic (CPV) systems focus light onto photovoltaic (PV) cells. Stated differently, they entrain a large solar radiation area onto a small cell, which functions at an irradiation level many times higher than that of unimpeded, direct sunlight. A PV cell's conversion efficiency actually increases somewhat with increasing irradiation levels. It can produce significantly more power when used in a concentrated environment as opposed to direct sunlight. By spreading out the cost of the cell and related optics, housings, and tracking systems over the high energy output, CPV technology takes advantage of this to drastically lower the cost of energy. Concentration ratios, which are commonly expressed in "suns" and are, although there are various definitions, generally defined as the ratio of the irradiance on the cell to the irradiance at the concentrator's entrance aperture. High-concentration photovoltaic (HCPV) devices have been constructed between 250 and 1,700 suns, while low-concentration photovoltaic (LCPV) devices function between 1.25 and roughly 40 suns. As will be shown later, the economics of devices with concentration levels in between HCPV and LCPV, also known as medium concentration photovoltaic devices, or miniature concentrating photovoltaic devices (MPCV), have not attracted much attention [1, 2].

Any of the CSP concentrator system types—trough, tower, linear, Fresnel, or dish—that have been covered in previous chapters can be used with CPV. However, the concentrators are made especially for the CPV application. CPV is also approached using Fresnel lenses.

6.1.1 Historical Summary

Despite years of development, commercialization of CPV technology has proven elusive due to technical and reliability issues that have hindered the development

Concentrated Solar Power Systems, First Edition. Bellamkonda Pragathi and D. P. Kothari.

of this seemingly straightforward concept. Significant barriers to market entry were also raised by the rapidly maturing silicon panel market, which had a head start many decades earlier. More information about the predicament the CPV industry faces, which also explains why the anticipated rise in commercial investment did not materialize during that time.

However, developments in the efficiency of useful high-performance multijunction cells, initially reported by US National Renewable Energy Laboratories (NREL) scientists Olson and Kurtz in 1994, have reignited interest in HCPV during the last 10 years. These cells promise conversion efficiencies of well over 40%, but at a cost that makes them unfeasible for use in standard panels—they were initially created for the space applications market and use materials other than silicon. These cells can only be used in an HCPV system in the terrestrial environment. It's important to note that this revival happened during a period when there was a lot of interest in energy costs and sustainable practices, primarily from venture capitalists looking to capitalize on the post-internet boom market. Furthermore, by the time the investment industry began to examine CPV with any seriousness, these cells had accumulated millions of successful cell-hours of operation and had passed the space industry's strict reliability standards. The investment sector gained confidence in the early years of the new century due to the convergence of performance promise, interest in renewable energy sources, and positive reliability data. As a result, the number of HCPV companies increased dramatically. Building on the groundbreaking work at NREL, significant progress had been made on new cell morphologies and related technologies, prompting investments into the cell segment itself shortly after [1, 3].

Furthermore, there was a lot of interest in the LCPV segment. Even though they weren't as noteworthy as multijunction cell advancements, silicon cell advancements were nonetheless significant for this industry sector. Designs that made financial sense were made possible by the combination of low fabrication costs and efficiencies in the range of 18%. LCPV is a smaller market than HCPV, but it has drawn excellent commercial representation.

Although the physics of reflection, refraction, and total internal reflection (TIR) limit all concentrator optics, Welford and Winston's [4] relatively new field of nonimaging optics (NIO) provided additional room for creativity within these constraints. Applying NIO has resulted in significant gains in performance and manufacturability, and it has made the path to useful, deliverable products much more pliable. Because of this, numerous new CPV businesses have attempted to combine NIO with the new cells. Additionally, a wide variety of designs have lately surfaced at both ends of the concentration range, adding to the already well-prototyped field.

The top CPV companies today have developed their products to a mature state, put high-volume production lines into operation, and gathered vast amounts of data from active installations. Several of these businesses are now concentrating on demonstrating their bankability and product dependability as more substantial business opportunities present themselves [1, 5].

There are currently more than 20 active CPV businesses, and Table 6.1 summarizes the top players. Note that a sample of product photos and a more thorough explanation of the methods employed by many of these businesses can be found later in this chapter.

CPV is beginning to live up to its long gestation period. CPV is "a long range option of vital importance to the global energy security." Cost analyses show that it most definitely has the potential to emerge as the PV approach with the lowest possible cost in large installations. The long-term period is approaching.

Table 6.1 Photovoltaic concentrator manufacturers in or close to production as of August 2010.

Company	System	Installations (completed or under construction not announced)
SolFocus	High-concentration, multijunction cell, glass, point-focus reflective optical array	8 MW
Amonix	High-concentration, silicon cell, acrylic, point-focus refractive (Fresnel lens) array	9 M
Amonix	High-concentration, multijunction cell, acrylic, point-focus refractive (Fresnel lens) array	0.48 MW
Entech	Medium concentration, silicon cell, single acrylic line focus refractive (Fresnel lens) module	0.2 MW
Concentrix	High-concentration, multijunction cell, glass/silicone point-focus refractive (Fresnel lens) array	1.7 MW
Opel	High-concentration, multijunction cell, point-focus refractive (Fresnel lens) array	0.44 MW
Solar Systems/Silex	High-concentration, multijunction cell, large single piece glass point-focus reflector (parabolic) system	1 MW
Emcore	High-concentration, multijunction cell glass/silicone, point-focus refractive (Fresnel lens) array	1.0 MW
Skyline Solar	Low-concentration, silicon cell, large single piece aluminum line focus reflective system (DSMTS system, described later)	0.1 kW

6.2 Fundamental Characteristics of Concentrating Photovoltaic (CPV) Systems

It is vital to comprehend the technology's features in order to appreciate its allure and the potential addition it may make to the already extensive array of solar technologies. This section discusses a few essential ideas.

6.2.1 Acceptance Angle

A CPV system can be compared to a telescope that is positioned in front of an efficient PV cell, regardless of the concentration level or optical technique employed. There is a restriction not present in standard flat plates or one-sun telescopes: only light entering this telescope will reach the cell and be converted to electricity.

A concentrator system possesses a field of view or *acceptance angle*, which is inextricably linked to the concentration ratio. For the general rotationally symmetric concentrator, the relationship for the maximum achievable geometric concentration ratio is derived by Welford and Winston [4] to be

$$C_g = \left(\frac{n_{out} \sin \emptyset}{n_{in} \sin \theta}\right)^2 \tag{6.1}$$

where C_g is the maximum possible concentration ratio, n_{out} is the index of refraction of the output medium of the concentrator, $\emptyset$ is the θ half-angle of the edge ray emerging from the concentrator output, n_{in} is the index of refraction of the input medium of the concentrator, and θ is the half-angle of the edge ray entering the concentrator: the acceptance angle.

Rearranging this equation for acceptance angle gives

$$\theta = \sin^{-1}\left(\frac{1}{n_{in}}\sqrt{\frac{(n_{out} \sin \emptyset)^2}{C_g}}\right) \tag{6.2}$$

Based on inspection, the concentrator's exit aperture at $\emptyset = 90^{\circ}$ would allow the output rays to radiate over the whole hemisphere, which is the maximum theoretically achievable acceptance angle. An acceptance angle of $\pm 2.7^{\circ}$ would be achieved by a concentrator with 1,000 suns operating in air ($n \geqslant 1$) and a low-iron glass final concentration stage (about 1.5). In actuality, though, this figure is unachievable since PV cells show an approximate cosine relationship between input angle and conversion efficiency (Spectrolab 2010). This is because the high-angle light would not be used effectively. The acceptance angle decreases to $\pm 1.2^{\circ}$ if the maximum allowable output half-angle is assumed to be at the location where the cell output power is 90%, or $\pm 26^{\circ}$.

As a result, the majority of solar radiation that CPV systems detect is the direct component of solar radiation, which comes straight from the sun and is not reflected, refracted, or scattered by the atmosphere or other terrestrial objects. Direct irradiance is the term used to describe the direct or beam power. In actuality, a design becomes more costly and potentially less efficient as it gets closer to the practical limit for acceptance angle. The intriguing and challenging problem of balancing efficiency, cost, manufacturability, and acceptance angle is at the heart of all concentrator design [6, 7].

6.2.2 Principles of Photovoltaic Devices

A PV cell is used in CPV systems to convert light into electricity, and their electrical properties are similar to those of conventional, flat silicon solar panels. Even though this chapter cannot adequately address the physics of PVs, it is still necessary to provide an overview of the subject. Green [8] provides a thorough development in a very succinct manner.

PV devices are made of semiconductor materials that directly convert light into electricity. These devices are constructed from materials that are intentionally "doped" with impurities to add more weakly bonded electrons to the crystal. These electrons reside in the valence band, which is a range or band of energies when they are not excited. One of these electrons can enter the conduction band, a higher energy band, if it gains enough energy from something like a photon interaction. The crystalline material's bandgap determines the amount by which energies in the conduction band are greater than those permitted in the valence band. Electrons can effectively jump over a forbidden energy zone, represented by the bandgap. Temperature and the materials—silicon, gallium arsenide, and germanium, to mention a few—are the main factors that determine the bandgap. Electron volts (eV), which measure the energy required to move one electron through a volt of potential, are used to express bandgaps. A photon needs to have energy at least equal to the bandgap in order to excite an electron from the valence to the conduction band. Any excess photon energy that reaches the bandgap is lost as heat. Excited electrons are very mobile once they are in the conduction band and can flow from the material to an electrode to create a DC electric current.

The structure created by combining two types of semiconductors—n-type and p-type—into a diode arrangement—commonly referred to as a photodiode when used to generate electricity—is essential to the extraction of electrons from the crystal lattice. In the absence of this, any liberated electrons would rapidly return to the valence band and release their energy in the form of radiation or heat through various intricate processes collectively referred to as recombination. Although recombination occurs throughout the crystal lattice, it is most intense near the surface. Recombination can be reduced, but not completely eliminated, by the diode's design and production methods.

The energy of a photon is inversely related to its wavelength. In general, photons of blue light are absorbed closer to the surface of a semiconductor due to their higher energy compared to photons of red light, which have wavelengths between 620 and 750 nm. The excess energy that photons with energies above the bandgap contain is lost thermally and can be converted to an electric current. Device conversion efficiency drops as photon energy increases, or wavelength decreases, from the minimum value needed to activate a bound electron. This is due to the fact that highly energetic photons are more susceptible to surface recombination [9, 10].

Similar to this, photons with energies lower than the bandgap energy in the solar spectrum are unable to move an electron from the valence band into the conduction band. Rather, the photons travel through the material and ultimately become adsorbed in a significant proportion, converting into waste heat through various undesired processes that are intrinsic to semiconductor materials.

Combining the aforementioned two mechanisms will result in a peak wavelength for the device's conversion efficiency and sensitivity only within a narrow range of wavelengths, depending on the characteristic bandgap energy. According to Kurtz and Geisz [11], the ideal bandgap for the solar spectrum is roughly 1.4 eV. The close bandgap of monocrystalline silicon (c-Si) is one of the reasons silicon has dominated the PV industry up to this point. Figure 6.1 shows the properties of silicon by plotting the efficiency against the wavelength of light.

The Solar spectrum with an AM (Air Mass) of 1.5 is a standard spectrum, produced when the sun is 48° above the horizon (a reasonable average estimate for the entire day), and is the result of modifying the sun's native spectrum by

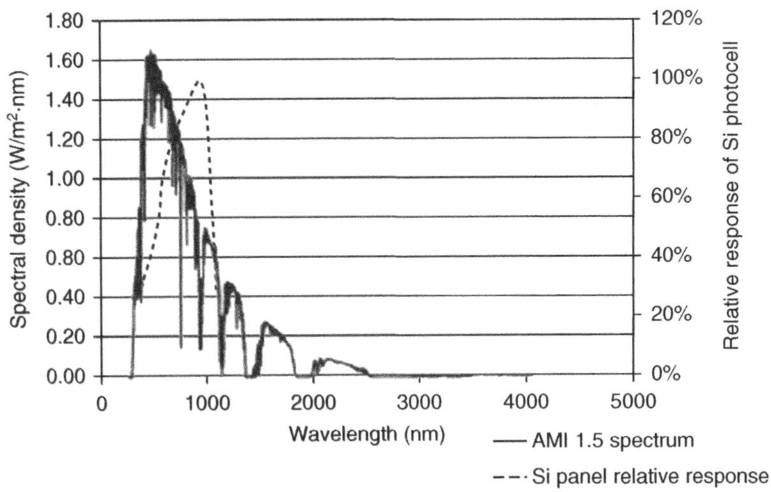

Figure 6.1 AMI 1.5 spectrum, overlaid with silicon junction conversion relative efficiency plot.

1.5 times the standard atmospheric depth. It is evident that silicon's properties restrict the amount of solar radiation that can be converted to electricity—not all of it—resulting in the upper bound efficiency. According to Green [8], this limit is roughly 27%.

The photodiode can be modeled at the device level using the equivalent circuit of Figure 6.2, and the solid curve in Figure 6.3 (also referred to as an IV curve) illustrates the relationship between voltage and current. A power–voltage curve is produced by multiplying the current by the voltage.

The broken curve in this figure highlights a crucial component of using these devices: in order for them to generate their maximum output, they must run at the proper voltage, also known as the maximum power point voltage, or V_{mp}. PV power management systems, such as inverters, are always looking for this peak power [12].

A figure of merit known as the fill factor has been developed as a single-valued "quality factor" to account for the deviation from ideal photodiode characteristics

Figure 6.2 Single-junction solar cell, equivalent circuit.

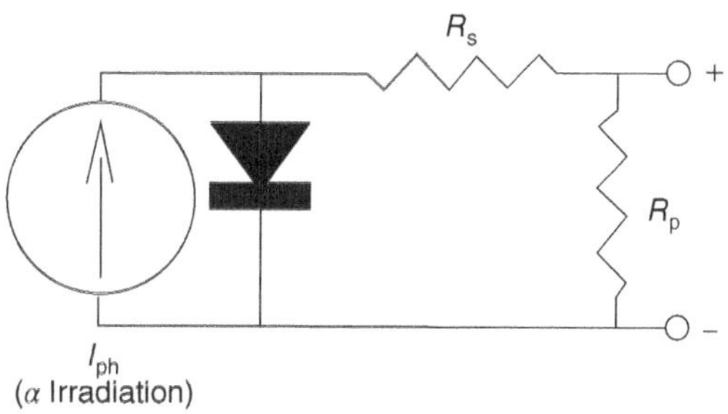

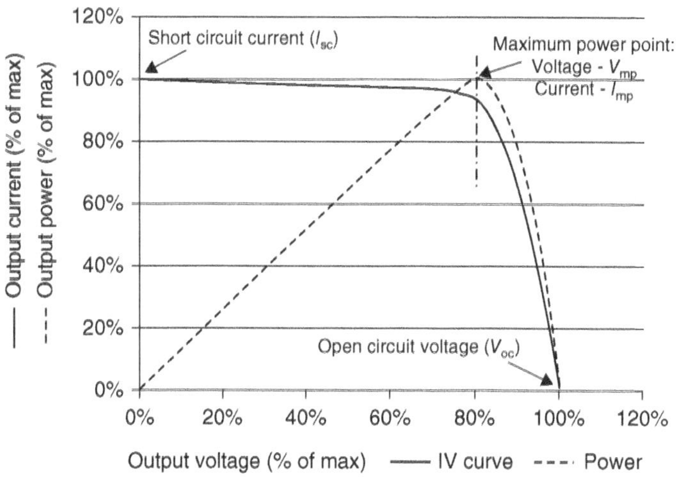

Figure 6.3 IV and power curves, photovoltaic junction under illumination.

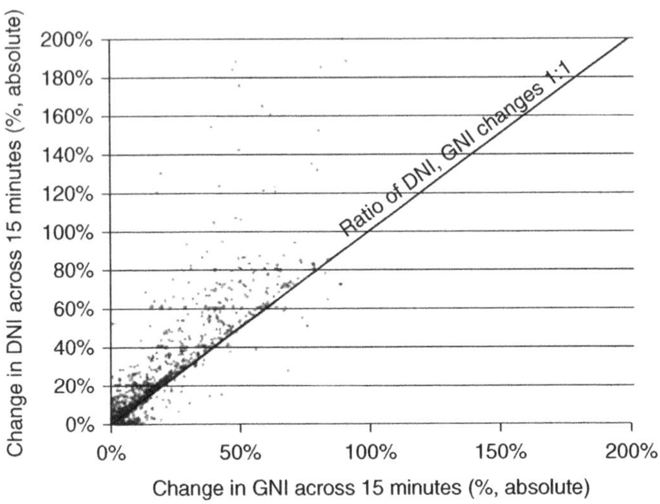

Figure 6.4 Fill factor.

caused by parasitic series and parallel resistances (R_s and R_p in Figure 6.4). The higher the fill factor value, the "squarer" the IV curve is and the closer the device is to ideal. This is equal to the product of V_{mp} and I_{mp}, divided by the product of V_{oc} and I_{sc}. Because the ideality factor of the diode itself contributes to the rounding of the IV curve, the fill factor in practice cannot reach the maximum of 100%, even for photodiodes with very low R_s and very high R_p. Different leakage mechanisms in the p/n junction cause the departure from ideality; some are due to impurities and imperfections in the crystalline material, and others are inherent to semiconductors operating above absolute zero.

Although the specific cells used in concentrator systems have slightly higher fill factors than silicon, concentrator systems' output characteristics are very similar to those of standard PV systems, as was previously mentioned. Most of the time, standard inverters are used, and recognizable stringing systems are utilized. Concentrator outputs are more variable due to the reduced acceptance angle when compared to flat panels because direct irradiance varies more than global irradiance. Data shown in Figure 6.3 demonstrate this, with a plot of the change in the direct irradiance (direct normal insolation, DNI) and Comparable shift in the world's irradiance (GNI). The direct component is impacted significantly more than the other component for the great majority of changes in the global value.

Concentrators lack intrinsic storage or hold-over capacity, just like any other solar system. They only generate electricity when the sun is shining, and other generation assets, like the grid or a nearby generating plant, usually support their use. However, very intriguing results have recently been obtained from intensive research into large-scale electrochemical storage, and in a few years, economically feasible direct electrical storage is predicted to greatly alleviate this constraint. The reader is recommended to start with Baxter [13] for an introduction to this rich field.

6.2.3 Maintenance

Cleaning the optical surfaces of a concentrator that are exposed to the environment, like the mirrors in an exposed mirror system, is the most frequent maintenance task. This is because, in comparison to unconcentrated systems, they will be more sensitive to light scattered by dust deposited on their surfaces due to their limited acceptance angle. Although this varies greatly depending on the site, HCPV companies have found that four times a year on average is sufficient; for LCPV, this will likely be less. Compared to systems where the external optical element is transmissive, like a window, exposed mirrors will have a higher sensitivity. Since passive air cooling is used in the great majority of CPV systems, water is only needed for cleaning. Further reducing the amount of water used during operation is the ability to mechanically clean systems that present a flat, glass face to the atmosphere using a “squeegee,” a standard window cleaning tool based on a sponge and rubber wiper blade (as opposed to being deluge-washed). Additionally, the requirements for water purity with a squeegee are not nearly as strict as they are for deluge, where deposits from evaporation must be routinely removed using alternative methods. One HCPV equipment manufacturer, SolFocus, estimates that their cleaning requirements—in situations where windows are the first optical surface—are less than 15 l of water per megawatt-hour.

The tracker is mostly responsible for the balance of maintenance. Regular maintenance for mechanical systems should only involve a biannual lubrication if they are well designed [14].

6.2.4 Energy Payback and Recyclability

As CPV systems are being used to offset dirtier sources of energy, it is important to understand exactly how clean they are. CPV systems stand apart from other forms of solar in two distinct ways. First, they are predominantly assemblies of optics and support mechanisms, with very small amounts of active PV material. Per unit area, the amount of energy used to fabricate the cells (whether silicon or multijunction) dominates over the manufacture of other materials, such as glass, aluminum, and steel. For CPV, the small area of cell material compared to the area of system will drive down energy payback time relative to standard PV. Reich-Weiser et al. [15] calculated the energy payback time for the SolFocus SF-1100 concentrator panel at 0.7 years, for example, against 2.2–2.7 years for silicon and 2.2–4.9 years for CSP under equivalent analysis. Second, again because concentrators are predominantly an assembly of common materials, they are highly recyclable, and over 95% of a typical system can reenter the manufacturing stream at the end of life.

6.3 HCPV-Specific Characteristics

Since only point-focus systems can achieve high-concentration ratios, the acceptance angle will essentially remain constant around the optical axis. This implies that all HCPV systems need to track the sun in both azimuth and elevation; the tracking accuracy increases with decreasing acceptance angle. Therefore, an HCPV system's acceptance angle is a crucially important figure of merit. Furthermore, it is only through the application of multiple-element optical systems and the science of NIO that feasible acceptance angles can be achieved. The accuracy needed for single element optics and classical designs, like those on some central tower solar thermal plants, will be two to four times that of a well-designed NIO system.

The energy that an HCPV system harvests is also influenced by acceptance angle; this energy is typically expressed as the kWh/kWp ratio, or the generated energy to peak power. Circumsolar radiation, which is measured from the sun's limb out to about 10° from the solar axis, can be significant under nonideal conditions for HCPV. According to Thomalla et al. [16], circumsolar radiation can reach several percent of the sun's irradiance at 0.5° from the axis when high cirrus clouds are present. In these weather conditions, energy harvesting will be lost if the circumsolar irradiation is not within the acceptance angle. HCPV systems are generally thought to be sensitive to beam or direct radiation only [17]; however, in reality, all of them have some circumsolar performance. Direct normal insolation, or DNI, is the term used to describe the beam energy. It is typically expressed in $kWh/m^2/day$ [18].

6.3.1 Two-Axis Tracking

Given that HCPV systems must track the sun in two axes, they do not suffer from 'cosine loss', the decrease in output that comes from energy striking the cell at large angles from the normal. This effect decreases the energy harvest from fixed tilt flat panels, with power production approximating a cosine output with the maximum centered local solar noon. CPV, on the other hand, has a broad daily output characteristic, which is extremely useful toward the end of the day, as it

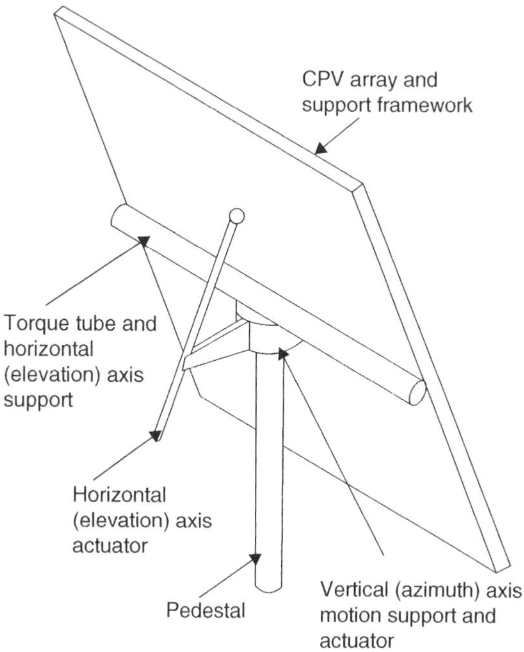

Figure 6.5 Pedestal or azimuth-elevation tracker.

is usually coincident with some of the highest daily electricity demand times. To counteract this, there are examples of silicon panels mounted on two-axis trackers, but their lower output has made this a difficult economic proposition. More commonly, the more efficient silicon panels are effectively mounted on single-axis trackers, with a fixed latitude tilt.

Where pedestal style trackers are used (Figure 6.5), their shape and spacing guarantee that there is no place on the field that is permanently shaded. This has become an important issue with permitting authorities, many of whom are concerned with minimal land disturbance, and whether native vegetation can be reestablished after construction. Linear trackers, used in both the CPV and concentrating solar power (CSP) industries.

In some places, stationary rack-mounted systems will create permanent shade, and they will significantly disrupt the local ecosystem.

One drawback of two-axis tracking is that it requires systems to be placed farther apart than stationary installations in order to reduce shading by neighboring systems in the early morning and late evening. Ground cover ratios, or the proportion of PV's total area to the land area on which it is installed, typically range from 17% to 22%. Due to high CPV output, this leads to less than the 40–50% found in stationary systems, but it still produces very competitive land use [19, 20].

6.3.2 Multijunction Cells

Multijunction cells are a recently developed technology that HCPV systems can uniquely take advantage of. Below is a summary of these cells.

Numerous semiconductor materials are photoactive, as was previously mentioned, and when they are used to create photodiodes, or "junctions," they can generate useful power. The semiconductor's bandgap ultimately defines the maximum efficiency by dictating the portion of the solar spectrum that the cell is sensitive to.

Research into a stack of electrically series-connected photodiodes, where each junction or photodiode is made of a different semiconductor material, has been ongoing for a number of years in an effort to break the barrier of single-photodiode efficiencies. With the largest bandgap material, the top junction is the one that light sees first and is capable of absorbing and converting only the highest energy photons to power. Lower energy photons are transferred from the first junction to the second, which is made of a lower bandgap material. Until the entire solar spectrum is covered, this structure is repeated. The total power is equal to the total power generated by all of the junctions if they are constructed so that their photocurrents are the same. For stacks of this kind, theoretical research indicates a fundamental efficiency limit of about 65%.

Currently, practical multijunction cells have three junctions, and the materials used in the various designs are remarkably similar, with a few notable exceptions. The middle junction is likewise epitaxially constructed and is made of GaAs with a bandgap of 1.4–1.4 eV. The first junction is constructed epitaxially and is made of InGaP with a bandgap of 1.8–1.9 eV. With a bandgap of 0.7 eV, the bottom junction is typically diffused into Germanium. From 25% in 1990 to 42% in 2010, efficiencies have increased significantly in the last few years. New research on cells with more than three junctions is starting to produce results above 40%.

Additionally, the temperature at which they operate has little bearing on their efficiency. In contrast to silicon cells, which have coefficients of approximately 0.4%/C, contemporary multijunction cells can have values as high as 0.07%/C. Because of this, a concentrator's cooling system is primarily made to keep the cell operating within its designated temperature range rather than to maximize output. When the ambient temperature rises, CPV fields exhibit minimal output degradation. They are particularly appealing from an economic standpoint in hot, dry desert conditions, where daytime temperatures can surpass 50°C.

But nothing is free, and any one of these intricate cells has a number of inherent problems that need to be resolved. A parasitic diode will first form between the various junctions if they are stacked on top of one another. The elimination of these parasitic diodes is imperative as they will significantly reduce the overall structure's performance. To isolate each individual photodiode, tunnel diodes are inserted in between the main junctions [21].

Second, every junction functions as a current source when it is properly biased at its maximum power point. They must be sized to produce identical photocurrents due to this and their series electrical connection, since the stack's total current will equal the junction with the least amount of current generation. Because of this, they are more spectrally sensitive than single-junction cells. Additionally, because the sun's spectrum varies throughout the day, one or more of the junctions will inevitably perform below par most of the time, dictating the photocurrent of the device. As a result, well-designed CPV systems will show a few percent daily efficiency variation. The concentrator optical transfer function requires careful consideration during design since it alters the incoming energy spectrum. A noteworthy development in this regard is the successful investigation of the bandgap-widening effect of nanostructures, such as quantum dots or quantum wells, which will reduce the spectral sensitivity of the cell and boost overall energy harvest.

These two worries create a very complex device with lengthy technical development times and, ultimately, a high intrinsic product cost, combined with other practical crystallography issues. A multijunction cell produces a lot of output, which only HCPV systems can use economically.

It is important to emphasize one last point, which was covered in passing above. Even after years of development and steady performance in space, multijunction cells are still in their technical infancy and have potential for further development. At 25°C, the efficiency of commercially available cells is currently about 48.5%, and some companies are demonstrating cells at about 42%. Despite the ghostly problems outlined above, performance can be greatly increased, and a number of respectable companies have three- to five-year product roadmaps that are nearly 50% efficient. The multijunction cell has, in fact, demonstrated consistent, deliberate performance improvement every year. Increasing efficiency has a significant impact on how much HCPV systems cost. The cost of the concentrators themselves drops for a given power output, and balance of plant and operating and maintenance (O&M) costs go down as well because there are fewer systems to install per megawatt.

6.4 LCPV-Specific Characteristics

LCPV systems, as previously noted, are found in concentration ratios ranging from 1.25 to 40. Due to their low concentration, they can employ designs such as parabolic troughs, linear Fresnel reflectors, the compound parabolic concentrator (CPC) or the V-trough that have lower achievable maximum concentration ratios. It is possible to think of these two-dimensional concentrators as concentrating in one axis only, with the other axis functioning at one sun. Furthermore, in comparison to HCPV systems, their acceptance angle in the concentration axis is larger.

The acceptance angle in the constrained direction will be above 24.75°, which is the tropical latitude plus the sun's angular radius, if the concentration is very low, say, about 2–2.5 suns. The system won't need to track if the concentrator is placed carefully so that the sun's arc falls within the acceptance angle throughout the year. See Luque (1989: 405) for a theoretical treatment. These systems are typically built as compound parabolic troughs or V-troughs for economic reasons [22].

Concentrators can be tracked by making seasonal adjustments alone as concentration rises from 2.5 to, say, 10 suns. They are moved twice a year so that the sun's arc lies within their acceptance angle of up to 12° for the ensuing six months. Moving forward from 10 suns, the tracking specifications get more demanding while still requiring movement in a single axis.

The LCPV system's ability to function with circumsolar and, in very low-concentration systems, the majority of diffuse light is another effect of the high acceptance angle. Because of this, in conditions with intermittent cloud cover, an LCPV system will respond to output signals more smoothly than an HCPV system.

Cost is a major obstacle to the design, manufacture, and operation of an LCPV system for all of the reasons mentioned above. Since high-performance multijunction cells are more expensive than LCPV systems due to the low-concentration levels used, silicon cells are used instead. As a result of the decreased output, very cheap optics, heat management, and tracking systems must be used because there is less power available to amortize the support structures and mechanisms. Any low-cost PV technology, such as thin film, could theoretically be used; however, high-efficiency mono-crystalline silicon has been required to achieve feasible optics and tracker costs. Additionally, the amount of effort that can be put into system maintenance is limited due to the lower output, meaning that extremely low-cost systems also need to be highly dependable.

6.5 Medium Concentration Photovoltaic Devices (MCPV)

There is sufficient proof that viable products can be developed and run, as evidenced by the interest displayed at both ends of the concentration scale. As of yet, there is more evidence of differentiation into market segments than of HCPV dominating LCPV or vice versa. It's interesting to note, though, that equipment with concentration ratios between 50 and 200 doesn't seem to be receiving much attention.

Parabolic troughs can function above 40 suns at the lowest end, but NIO predicts that their theoretical maximum concentration will be approximately 100 suns. Once they reach this limit, tolerances and accuracy requirements make them costly, and their output is too low for efficient amortization. The benefits of troughs in terms of mechanics and hydraulics have led to the refinement of designs in the solar thermal space, with a practical and cost limit still well below theoretical limits. This has produced designs with about 80 suns.

Dual-axis tracking and multijunction cells are standard at the top end of the spectrum. Systems still require trackers as concentration drops below, say, 250 suns, but the cost per unit power rises due to the decreased power density. The high cost of the cell must eventually be justified, at which point silicon must be used instead. However, the two-axis tracker is not justified given the reduced output. For a few years, a concentrator company used silicon to run a point-focus system at about 150 suns. However, due to financial constraints, they had to increase the concentration and switch to a triple-junction cell.

In conclusion, cost optimization is primarily responsible for both ends of the concentration landscape's occupancy. While LCPV systems rely on less strict requirements for their optics and trackers, which result in lower cost, HCPV systems have sufficient output over which to amortize their relatively complex designs [23].

6.5.1 Application to the Market

The market for viable solar covers a huge range of regions and weather patterns, and it is made considerably more difficult by the lack of access to financing, construction resources, transmission assets, and power users/offtakers. An additional and significant complication is added by semigovernment organizations like utilities and all tiers of government. Furthermore, interest in large installations has increased despite the fact that for many years the residential rooftop market dominated the growth of the solar sector due to the reliability of PVs and their increasingly attractive costs. Solar power plants with a capacity of up to hundreds of megawatts are currently being developed and should be operational well before the end of the decade. These plants are typically ground-mounted and are found more and more in hot, dry regions of the world (such as the Mediterranean). As a result, there are numerous competitive ecosystems that support different technologies that are currently in use. Where are concentrators appropriate? is the question.

The nontracking, very low-concentration devices will be used in conjunction with conventional silicon panels. Typically, their economic arguments revolve around panels that use less silicon and have outputs comparable to standard panels, but at a lower cost. They have a lower acceptance angle than standard panels, which makes them more susceptible to light scatter. As a result, they require more frequent cleaning and have a restricted useful life in extremely hazy conditions. This will put pressure on bigger installations—like those on business rooftops—where the cost of operation can be spread across a greater number of panels. Low-concentration devices are also likely the only practical concentrator for rooftop applications because they don't require tracking. Higher concentration systems for rooftop applications have been developed by a small number of companies, but the economics have proven to be risky, particularly in urban environments with significant levels of atmospheric scattering. The stationary systems' comparatively low output makes them less competitive with higher concentration devices; therefore, field-mounted applications are unlikely to see much use for them.

Tracking LCPV systems, or those with 10 to 40 suns, are less susceptible to manufacturing tolerances than HCPV systems, but their output density is lower. Even though they produce more than stationary concentrators, their mechanical complexity and upkeep needs will likely prevent them from being installed on rooftops. However, their wide acceptance angles and high output levels will make them perfect for distributed generation in an urban setting. Commercial pressure will be put on installing tracking LCPV systems in fields of at least tens to hundreds of kilowatts due to maintenance requirements and the difficulty of installing mechanical systems. This will allow O&M tasks to be completed efficiently and economically. The high end of LCPV requires careful placement because the acceptance angle's small enough combined with the demand for extremely low manufacturing costs to put them at a disadvantage in high diffuse radiation areas. The best LCPV systems, like HCPV, will only be available in high DNI areas.

The highest power density of any solar technology is found in HCPV systems. They also have the lowest acceptance angle and necessitate the most sophisticated, costly tracking systems. When combined into power plants with capacities ranging from hundreds of kilowatts to several megawatts, HCPV equipment will be financially advantageous. These power plants will be operated in regions known as the "sun belt," which have high DNI. HCPV is particularly well-suited to hot, dry climates because of its high power output at elevated ambient temperatures combined with very little water consumption (and, in many cases, the ability to use untreated water). For CPV, this is a market niche that is sizable and expanding quickly. Based on market research, this represents about 20% of the overall market, or 50% of the ground-mount solar market [24, 25].

6.6 Design of Concentrating Photovoltaic (CPV) Systems

The preceding discussion on the characteristics of the technology and its entry points to the market now sets the stage for an introduction to the design elements and challenges of a concentrator. The basic issues at the heart of any CPV design are introduced in this section, and a small set of example products are described.

6.6.1 Levelized Cost of Energy

In the end, a CPV plant is a financial tool, and investors base their decisions about technology and project partners on return analysis. Furthermore, because a solar plant generates income over the course of its lifetime, any investor will need a precise evaluation of the plant's continuous financial performance in comparison to expectations. The levelized cost of energy (LCOE) is an efficient metric that meets both of these requirements. As with any CSP technology, LCOE was introduced in Chapter 2 and is a crucial metric for CPV system design [26]. In essence, it is a computation of the lifetime cost of energy for any generating plant, adjusted for inflation and discount rates, including all fixed and marginal costs, output degradation, projected repair and replacement costs, and net present value. This kind of calculation, which is expressed as ¢/kWh, is unique in that it can be used to compare technologies, such as CPV vs PV, hydroelectric, or fossil fuel generation. According to Short et al. [26], life cycle cost of energy (LCOE) is defined as "that cost that, when discounted back to the base year, will equal the total life cycle cost (TLCC) when assigned to every unit of energy produced (or saved) by the system over the analysis period".

The formal definition for LCOE, if tax issues are not considered, can be expressed as

$$\mathrm{LCOE} = \frac{\mathrm{NPV}_{\mathrm{LCC}}}{\sum_1^N (E_i/(1+\mathrm{DR})^i)} \tag{6.3}$$

where E_i is the energy output in year i, N is the amortization period, DR is the discount rate, and $\mathrm{NPV}_{\mathrm{LCC}}$ is the net present value of life cycle costs, which is given by

$$\mathrm{NPV}_{\mathrm{LCC}} = \sum_{i=1}^{N} \frac{C_i}{(1+\mathrm{DR})^i} \tag{6.4}$$

where C_i is a cost during period i.

The cost cashflows for a CPV system fall into one of the following categories:

- The price of installing the concentrator/CPV system, which will only require one upfront payment.
- In year zero, there will also be a single, sizable investment required for the installation of balance of systems (BOS), such as inverters and civil works.
- The annual cost of upkeep and operation.

It is possible to compare various technologies with varying financial structures and lifetimes at the time of a potential investment by bringing all costs and generation back to net present value.

The core idea of LCOE is the quantification of energy rather than operating power. When talking about solar plants of all kinds, installed capacity-centric costs ($/W or $/kWe) are frequently used. However, this approach has two significant disadvantages. First off, different technologies have different rating systems. According to Melia et al. [27], a 400 W HCPV panel will actually produce 400 W at 20°C and under 850 w/m^2 direct irradiance (the rating irradiance) due to the way the rating conditions are defined. However, the rating conditions for silicon panels differ, meaning that a poly-silicon panel with the same label will generate 264 W when operating at its rated condition of 1,000 w/m^2 global irradiance. These variations stem from test practicalities; however, if applied during the design stage, power plants of widely disparate sizes, with varying upfront costs and energy harvests, will be constructed.

Second, each technology has a different specific energy, which is a measure of how well it harvests energy. It is calculated as the annual ratio of kilowatt-hour to kilowatt rated. This can be attributed to various factors, such as the technology's mounting on a tracking system, the extent to which temperature affects output, the geographical separation of systems, the acceptance angle, and the technology's vulnerability to soiling. While a CPV system with a very narrow acceptance angle, for instance, might produce the same amount of power when pointed toward the sun as one with a wider one, the larger acceptance angle system will admit more circumsolar radiation over the course of the year and will be less prone to tracking errors, alignment problems, and foundation movement. More energy will be harvested by the product with a bigger acceptance angle, giving the investor a higher annuity. Thus, choosing technologies or project equipment suppliers becomes impossible when employing a power-based metric, which also provides very little information about the investment story.

The US Department of Energy and NREL have made considerable efforts to promote the usefulness of LCOE and provide resources for its application, which has led to its increasing acceptance as a standard metric. It is recommended that the reader look into the NREL Solar Advisor Model (SAM) and review the documentation that goes along with it.

6.7 General System Design Goals

A CPV system's overarching design objective is often to generate a product that minimizes the LCOE in the selected geography. That straightforward assertion, however, contains a number of difficulties and compromises. Scale is the first consideration.

Irrespective of their efficiency, PV technologies require extremely large-scale production and deployment. Imagine a CPV power plant rated at 40 MW Pdc, with equipment running at 27% efficiency and a 500 sun concentration ratio. To meet the output requirements at that efficiency, 108 W of input power would be needed.

1.4×10^5 m^2 of concentrator area would be needed at the 850 w/m^2 irradiance rating, which translates to 2.6×10^2 m^2 of cell at 500 suns. Concentrator cells typically measure 7.5 mm on one side, or 0.56 cm^2, so the plant would need to install 4.7 million of these cells. Electrical connections, cooling, alignment, a portion of the optical system, and ultimately integration on a tracker with practical dimensions are all necessary for each of these cells. It should be evident from the sheer volume of operations at this moderately sized plant that successful CPV deployment requires very high-volume manufacturing on the scale of the automotive sector, for example.

Fortunately, despite their apparent complexity, the majority of PV concentrator designs lend themselves exceptionally well to standard high-volume manufacturing techniques, which are associated with low production costs and are used, for example, in the automotive or electronics industries. Unlike thin film, for instance, where thousands of square meters of intricate vacuum-deposited material must be produced, they are essentially an assembly process.

Moreover, even with their intricacy, CPV systems need to function within recognized solar reliability standards, typically ensuring 80% output after 25 years. It is crucial to have a thorough understanding of the degradation mechanisms at play in any concentrator design in order to accurately measure them. Two things need to be taken into account, even though this might seem like a challenging task for a new technology. First, because concentrators are assemblies, the reliability of their individual parts and assembly techniques can be studied. Second, many well-designed subsystems employ materials, assembly procedures, and techniques borrowed from more seasoned industries. To ensure a dependable system, test procedures, design guidance, and degradation data are all available.

For example, an aluminum-drawn backpan for a concentrator can employ methods from the automotive sector, where materials selection, test procedures, and simulations can all ensure a 25-year lifespan.

Ultimately, performance is crucial, and the key idea is, as previously mentioned, minimizing the product's life cycle expenses. The financial performance of the field is determined by this metric, which is influenced by all other performance parameters.

Concentrator design offers a wide range of options, which has led to the emergence of numerous distinct product styles. There are actually so many that it has been nearly hard to classify them beyond HCPV and LCPV. To make matters more confusing, there are still a lot of undiscovered design possibilities. On the other hand, drawing necessarily imprecise generalizations from observing certain trends is possible. These broad categories will be explained in the remainder of this chapter. For illustration, a limited number of examples will be provided.

6.7.1 System Granularity

System granularity refers to the overall architecture, and whether the concentrator is composed of an array of small concentrators (usually each irradiating a single cell), or a single, large optic, powering an array of cells. There are instances of both approaches, covered in the examples below. An array of small concentrators is by far the most common approach, as it lends itself better to modern manufacturing techniques.

6.7.1.1 Optical Method

There are three types of optical methods: catadioptric (using both refractive and reflective techniques), reflective, and refractive. The most popular are the first two, with many people favoring refractive Fresnel systems. A decreasing number of companies are pursuing single reflective optics and cascadean reflectors. Catadioptric nonimaging optical elements are occasionally employed as a concentrator's final stage. Swanson [28] noted that Universidad Polytechnic Madrid designed and prototyped a complete concentrator of this type; however, it has not been commercialized. All other aspects of a concentrator design, including available manufacturing processes and thermal management, are influenced by the optical design.

6.7.1.2 Tracking Type

There are many mechanical styles of tracking systems. Degree of concentration will primarily dictate the selection of single- or double-axis tracking. There aren't many strong criteria after this. There seem to be benefits and drawbacks for each, but none that are particularly strong.

Elevation and azimuth tracking systems are mounted atop a pedestal or pole, with one axis rotating in a horizontal direction (elevation) and the other vertically (azimuth). A subset of these are polar tracking systems, in which the azimuth axis is typically inclined at the local latitude angle.

In tilt/roll systems, there is a single horizontal axis and an additional one that is attached to the first at a right angle. Figure 6.6 depicts an example of a tilt/roll tracker. Lastly, carousel trackers are a particular kind of azimuth/elevation tracker; however, they are incredibly small and may be appropriate for rooftop mounting. However, their mechanical complexity poses a challenge in terms of manufacturing and tolerancing.

Tracking systems can be steered or controlled in various ways. First, electric motors are typically used to power movement, though there are a number of hydraulically actuated systems available. The algorithms used by control systems can range from relying on an ephemeris equation to sampling the concentrator's output or using external sensors. Although there is still a lot of variation in the market, the most workable approach seems to be combining ephemeris with concentrator output monitoring.

The subject of trackers is complicated, particularly when factors like wind, regional building codes, structural vibrations, logistics, deployment, and lifetime are taken into account. Unfortunately, details are outside the purview of this one CPV chapter.

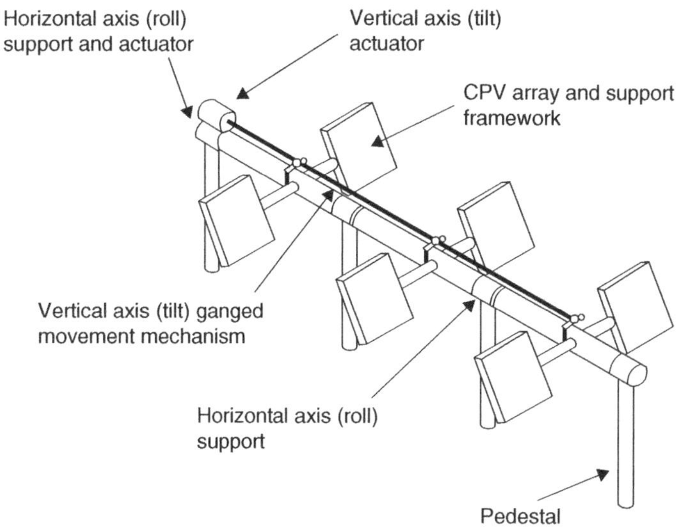

Figure 6.6 Tilt-roll tracker.

6.7.1.3 Environmental Control Methodology

Cell thermal management is necessary for CPV systems. Even with multijunction cells' high efficiencies, more than 50% of incident energy is still converted to heat, which needs to be eliminated to preserve dependability. The trend is generally toward dry, passive cooling, in which heat energy is conducted out of the cell and subsequently radiates and convects into the atmosphere. While some systems have experimented with active air cooling schemes, their implementation is typically hindered by the requirement for low maintenance costs and dependability. Larger optical systems and concentrators with extremely high-concentration ratios typically require active cooling, which can involve fluid reticulation and heat pipes.

Furthermore, shielding the optical systems from the weather is necessary. If Fresnel lenses were used outdoors, their delicate teeth and typically deep valleys would quickly lose their effectiveness. Almost always, they are housed in a housing with a flat front window and lens elements molded or embossed onto the inner, protective surface. Reflective systems are more flexible and come in a number of designs with external mirrors, much like solar thermal technologies. Nonetheless, a number of reflective technology businesses continue to decide to house the mirrors, so adding even more security.

Although active drying methods have been observed, passive air filtration is typically used to maintain the environment inside the housing.

6.7.1.4 Cell Administration

There are both financial and technical challenges in the immediate vicinity of the cell. Regardless of concentrator alignment, thermal management, strong electrical connections, mounting of a bypass diode, mechanical alignment, and potentially very high incident solar flux management must all be done within a few millimeters of the cell. Two general strategies have been used to address these numerous issues. One approach has been to mount the cell, or an assembly of the cell and bypass diode, on the concentrator superstructure and finish building the surrounding structure. In the second method, a separate assembly known as a "receiver" is made up of the cell or cell array and associated thermal, electrical, and optical components. This subassembly is then mounted to the concentrator in a separate manufacturing step.

6.8 Introduction: Relevance of Energy Storage for Concentrating Solar Power (CSP)

An overview of the state and recent advancements in storage technology meant for incorporation into CSP plants is given in this chapter. The thermal mass of the working fluid and of parts like absorbers, heat exchangers, and tubes in CSP plants already has an inherent storage capacity. By storing solar energy that is not immediately needed by the CSP plant's thermal process, storage units increase this capacity. The storage unit replaces all or part of the solar collector by supplying heat to the thermal process during discharge. One of the most notable benefits of CSP over other renewable energy technologies is the ability to integrate affordable local storage capacity. Unlike technologies like PVs, which must first generate electricity and then add the additional expense and efficiency loss of a complex independent electrical storage subsystem, storage can be integrated into CSP system designs in a way that delivers benefits with little to no impact on overall system efficiency and cost of energy.

Storage can benefit CSP in a number of ways. First, it can increase value by allowing electricity generation to be shifted to times when demand is high (Figure 6.7).

- Fossil-fired backup systems have high upfront costs, and their low thermal efficiency reduces the advantage of generating electricity from renewable energy sources.
- The capacity of external backup systems required to offset the mismatch between availability and demand can be reduced.
- Shorter start-up times by preheating absorber systems with energy from the storage system;
- Increased efficiency by preventing cloud-related transients in the power cycle.

Sioshansi and Denholm [29] describe the financial benefits of storage integration into CSP plants in the southwestern United States. The most significant finding is that, in general, storage increases the cost-effectiveness of CSP plants; however, the amount of improvement varies greatly based on the technology

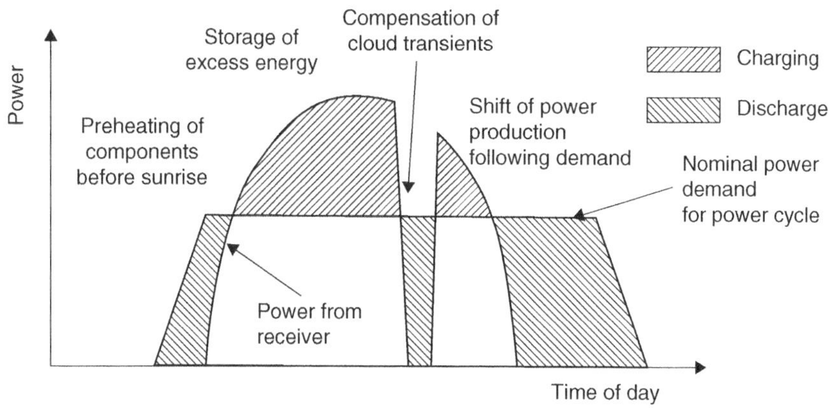

Figure 6.7 Various functions of thermal storage in a CSP plant.

and assumptions made for the particular project. This study also looks at the possibility of moving power generation to times when the outside temperature is lower, which would boost the power cycle's efficiency.

The majority of commercial thermal storage systems in use today, both for solar heating and industry, have storage capacities of less than 1 MWth and are operated at temperatures below 100°C. These systems differ from storage systems meant for CSP in two key ways: The temperature range required for CSP and solar process heat applications is between 120 and 1,000°C, which introduces unique requirements for material thermal stability and corrosion resistance. The enormous storage unit capacity is another feature of CSP applications. For a seven-hour operating period, a 50 MW_{el} parabolic trough power plant needs roughly 1 GWh_{th} of storage capacity. Physical limitations limit the energy density of thermal storage systems, so typical CSP storage systems need several tens of thousands of tons of storage.

It should go without saying that, within reasonable financial limits, a thermal energy storage system seeks to reduce the amount of energy lost during storage. However, the problem of temperature variations and the resulting loss of power generation potential must also be taken into account in systems that are being compared. All thermal systems need to be charged with a heat source that is hotter than they are in order to return heat to a working fluid that is colder than the system. Because power generation efficiency decreases with decreasing temperature inputs, the energy storage system's presence can either force lower temperatures, which will reduce power generation efficiency, or force higher temperatures, which will force solar receivers to operate at higher temperatures in order to make up for the loss of power.

Based on the physical idea behind heat storage, basic storage concepts can be divided into three primary categories. While sensible heat storage is used in today's commercial storage systems, research into high-temperature latent heat storage systems, which have now reached a high degree of maturity, has been spurred by the development of direct steam generation in the absorbers. The use of sorption processes and reversible chemical reactions makes up the third major group. This group is typically referred to as chemical energy storage. The different storage concepts that have been developed within the three main groups will be explained in the sections that follow.

6.8.1 Current Commercial Status of Storage Technology

Virtually from the start, storage system research has followed the development of CSP technology. A summary of storage systems incorporated into both commercial and experimental CSP plants can be found in Table 6.2. Storage capacity is already present in many commercial CSP plants today. A wide range of fundamental ideas, which represent the various absorber systems, heat transfer mediums, and power

Table 6.2 Survey of selected storage systems integrated in commercial and experimental CSP facilities.

Project	Concept	Concentrator type	Max temperature (°C)	Thermal capacity (MWh_{th})	Start of operation
Eurelios (Italy)	Two-tank molten salt, steam accumulator	Tower/heliostat	440	0.5	1981
SSPS (Spain)	Two-tank liquid sodium	Tower/heliostat	540	1.0	1981
NIO central receiver (Japan)	Steam accumulator	Tower/heliostat	250	4.0	1981
Solar One (USA)	Thermocline with thermal oil	Tower/heliostat	400	28	1982
CESA-1 (Spain)	Two-tank molten salt	Tower/heliostat	440	4	1984
Themis (France)	Two-tank thermal oil	Tower/heliostat	450	40	1984
	Two-tank molten salt				
SEGS-1 (USA)	Two-tank thermal oil	Trough	405	120	1984
TSA (Spain)	Packed bed with air	Tower/heliostat	700	1	1994
Solar Two (USA)	Two-tank molten salt	Tower/heliostat	565	110	1996
PS10 (Spain)	Steam accumulator	Tower/heliostat	245	20	2006
Andasol-1 (Spain)	Two-tank molten salt	Trough	485	1,010	2009
Solar Tres/Gemasolar (Spain)	Two-tank molten salt	Tower/heliostat	565	2,400	2011

cycles employed in solar thermal power plants, are central to the development of storage systems for CSP. A given application's unique boundary conditions, such as the working fluid, temperature range, storage capacity, power level, and reaction time, must be taken into consideration when determining the best concept.

6.8.1.1 Sensible Energy Storage

In sensible heat storage systems, variations of the stored energy ΔQ_{12} are dependent on the variation of the mean temperature (T) according to

$$\Delta Q_{12} = m \int_{T_1}^{T_2} c(T)\mathrm{d}T \tag{6.5}$$

where mass (m) and mass-specific heat capacity (c) are given. Physically, the specific heat capacity of the storage material and the available temperature differential T_1–T_2 limit the capacity of sensible heat storage systems. As an example, water has the highest specific heat capacity per mass of all liquids and solids, so for a temperature difference of 100 K, the maximum storage capacity for sensible heat storage systems is in the range of 0.11 $\mathrm{kWh_{th}/kg}$. Because of their low storage density, thermal storage systems require large masses in comparison to chemical energy sources (e.g., gasoline, 11.5 $\mathrm{kWh_{th}/kg}$).

The choice of storage material is the first step in the creation of a sensible heat storage system. Systems that use solid storage media are different from those that use liquid storage media. The idea used to transfer solar energy to the storage material is another factor used to categorize storage systems. The heat transfer fluid (HTF) that absorbs solar radiation doubles as a storage medium in direct storage systems. Systems for indirect storage employ a storage medium distinct from the fluid used for heat transfer. A classification of different concepts for sensible heat storage is presented in Figure 6.7. The ensuing sections go into further detail about these ideas.

6.9 Liquid Storage Media: Two-Tank Concept

Figure 6.8 shows the two-tank storage approach. As the name suggests, a liquid medium is circulated between a hot tank and a tank that is known as the "cold tank" even though it is still several hundred degrees. The different liquid media and properties necessary for thermal energy storage are displayed, Connecting the inlet and outlet of a heat transfer process to two distinct volumes maintained at different temperatures is the most straightforward method of storing sensible heat in a liquid. This method, which uses vertical single tanks with thermal stratification to separate the hot and cold volumes, is widely used for low-temperature applications utilizing water as the storage medium. A single vertical tank is typically not preferred over two separate storage tanks due to the large volume required for CSP applications. Direct storage of the HTF is an easy solution if the absorbers use a nonpressurized HTF.

Thermal oil was directly stored and ran between 240 and 407°C in the SEGS I parabolic trough power plant. Thermal oil is not thought to be a desirable option

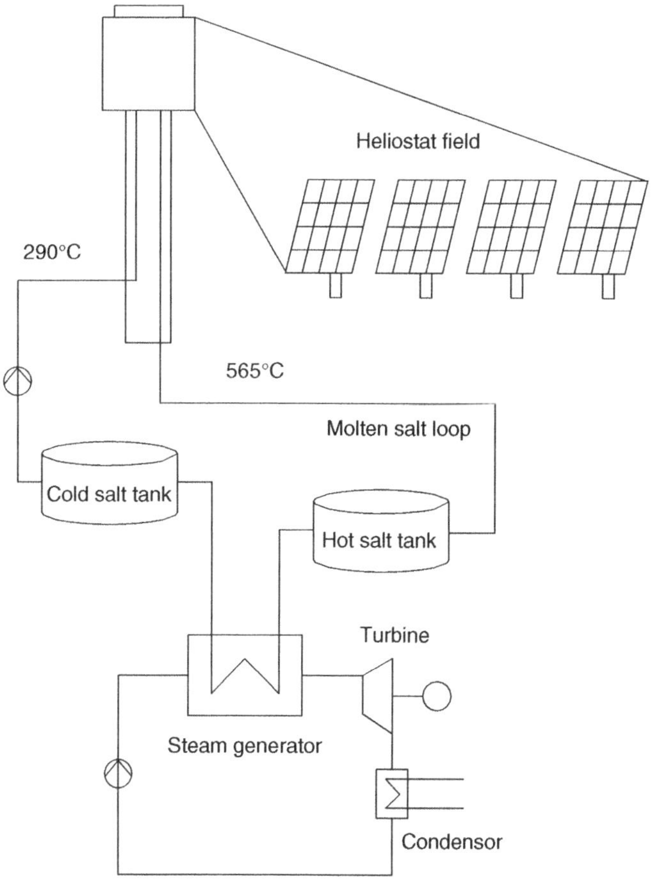

Figure 6.8 Simplified scheme of central receiver with two-tank storage concept.

because of its high cost, flammability, and restricted maximum temperature. An alternative that is already well known from other uses is molten salts. Thus far, nitrate mixtures have been the preferred method for storing energy in molten salts. The primary disadvantage of systems involving molten salt is the restricted range of operating temperatures, which is determined by the freezing point at the lowest point and the start of thermal degradation at the highest point. Since remelting is very complex due to the low heat conductivity, freezing within the storage tanks must be avoided. Because molten salts are more corrosive at higher temperatures, more expensive materials are needed for storage tanks exposed to higher temperatures. Using 1,400 tons of molten salt, the two-tank concept was successfully demonstrated as part of the Solar Two project.

A simplified design for a solar receiver that uses molten salt as a working fluid and storage medium is shown in Figure 6.8. Pacheco provides a thorough explanation of this storage system (2002). The 16-hour storage system of Torresol Energy's Gemasolar project also employs the direct two-tank concept (Figure 6.9).

Figure 6.9 Two-tank storage system of the 17 MW_{el} Gemasolar central receiver plant. Source: Torresol Energy, Spain.

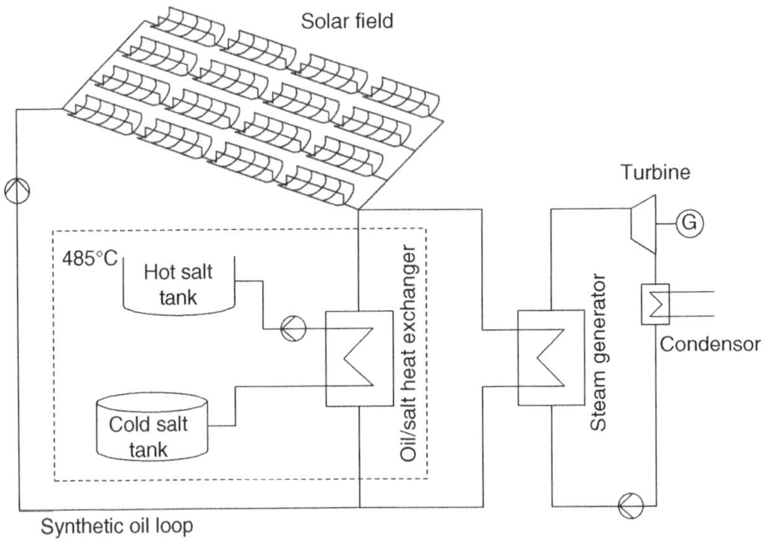

Figure 6.10 Simplified scheme of a parabolic trough plant using thermal oil as HTF with an indirect two-tank molten salt storage concept.

For example, the molten salt used as a storage medium in an indirect storage system differs from the HTF used in the absorbers. This idea is selected in cases where the costs of the HTF specifically exceed those of the storage medium (such as molten salt). A heat exchanger connects the individual loops. Using 28,500 tons of binary nitrate salt that is operated between 292 and 486°C, the first commercial two-tank molten salt storage unit integrated into the Andasol-1 plant is an indirect concept. Figure 6.10 illustrates the basic idea.

6.10 Liquid Storage Media: Steam Accumulator

Because of its compatibility, affordability, and high specific heat capacity, liquid water makes an appealing storage medium. Water used as a liquid storage medium needs to be pressurized at temperatures higher than 100°C. Saturated steam is provided during discharge by steam accumulators (Figure 6.11). Saturated liquid water under pressure is used as the source of energy to create saturated steam.

During the discharge process, the volume and pressure of the saturated steam produced by the storage system drop. In comparison to the mass of saturated steam supplied during the discharge process, there is a significant amount of saturated liquid water left inside the steam accumulator. Under the assumption that the steam accumulator contains a constant mass m liquid of saturated liquid water, the thermal energy supplied during the discharge from pressure P1 to pressure P2 can be calculated as follows:

$$\Delta Q_{12} = m_{\text{liquid}} c_{\text{liquid}} (T_{\text{sat}}(P_1) - T_{\text{sat}}(P_2)) \tag{6.6}$$

where T_{sat} is the pressure-dependent saturation temperature and c_{liquid} is the mean specific heat capacity of liquid water. Figure 6.12 shows the volume-specific

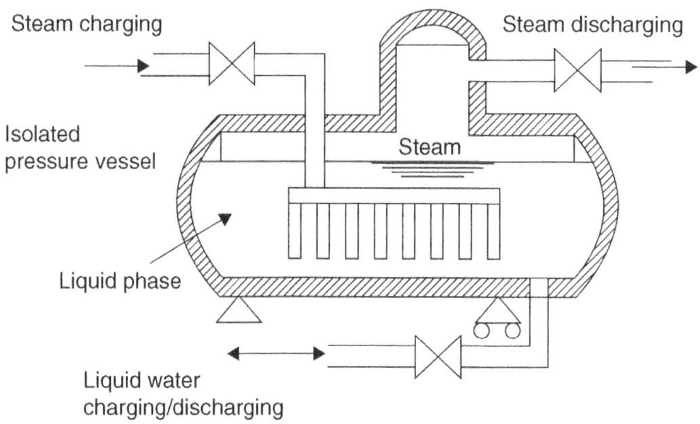

Figure 6.11 Scheme of a steam accumulator.

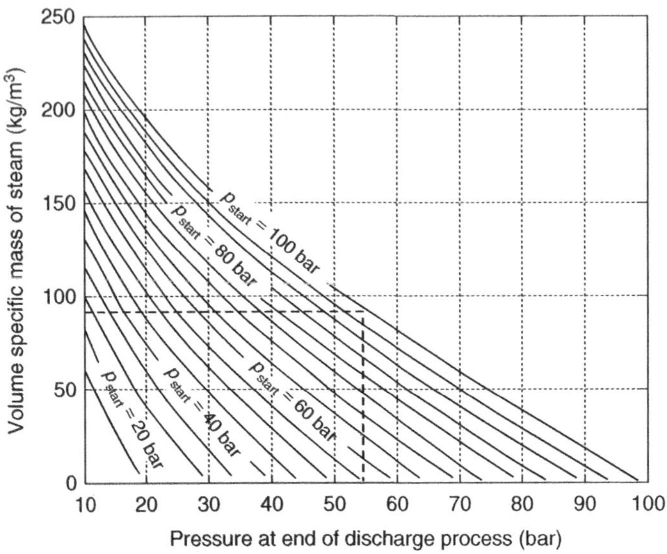

Figure 6.12 Volume-specific mass of saturated steam provided by steam accumulator for different initial pressures and pressure drops. Dashed line indicates an example with initial pressure 100 bar and final pressure 55 bar; steam accumulator delivers approximately 90 kg saturated steam per m^4 storage volume.

amount of steam provided by a steam accumulator dependent on initial pressure and pressure drops.

Steam accumulators are charged by feeding steam into the liquid volume. The temperature of the liquid volume is increased by condensation of the steam. While steam accumulators have fast reaction times, the storage capacity is usually limited economically by the costs of the pressure vessel. Since steam accumulators provide steam almost instantly for a short period, this concept can be used as buffer storage to compensate for short cloud transients.

Steam accumulators are widely used in the process industry in the temperature range between 100 and 200°C. The application of steam accumulators for large-scale CSP plants was described by Gilli and Beckmann [30]. Some experimental CSP plants such as Eurelios and the Japanese central receiver plant at NIO used steam accumulators. The PS10 central receiver power plant, representing Europe's first commercial CSP plant starting operation in 2007, uses steam accumulators.

For approximately 50 minutes, the 11 MW_{el} can operate at 50% load thanks to four tanks with a combined storage capacity of 20 MWh. The central receiver supplies saturated steam to the storage system at a pressure of 45 bar.

6.11 Solid Media Storage Concepts

Cost considerations are the primary driving force behind the use of solids as storage media (Figure 6.13). For example, the material costs for concrete are between 10% and 20% less than the corresponding costs for molten salt, and solid media storage systems should also have lower maintenance costs. Furthermore, there are no issues brought on by leaks, evaporation, or freezing. The primary obstacle to implementing a successful storage concept is the low thermal conductivities of cost-effective solid storage materials. Several solutions have been proposed to get around the storage material's heat transport constraints. Reducing the heat transfer path from the majority of the storage material to the transfer medium is the aim of these options.

6.12 Solid Media with Integrated Heat Exchanger

A parallel pipe tube register is typically integrated into the storage volume for pressurized HTFs. For this method, various castable storage materials have been researched, including concrete and castable ceramics. A test unit was connected to parabolic trough collectors at the Plataforma Solar de Almería in Spain (Figures 6.13 and 6.14). This system was intended to run on thermal oil at a maximum temperature of 490°C, with a thermal power of 450 kWth and a capacity of 400 kWh_{th}. To investigate the long-term behavior, a second test unit with a 400 kWh_{th} capacity intended for storage cycles of six to eight hours was connected to a test rig (Figure 6.15).

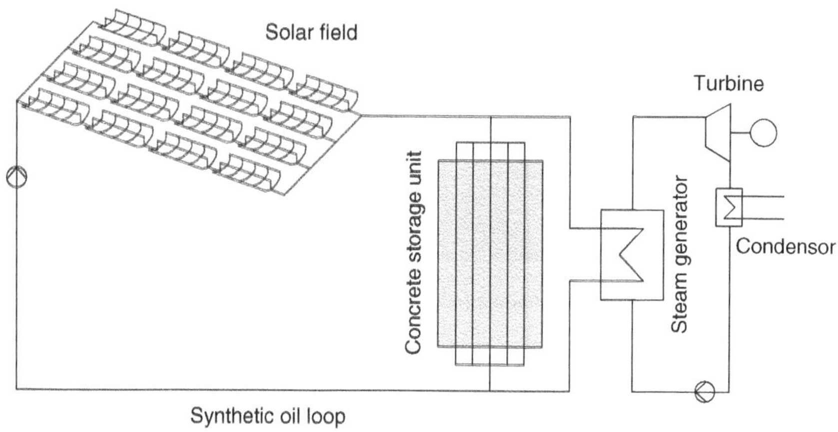

Figure 6.13 Simplified scheme of parabolic trough plant using thermal oil as HTF with a concrete storage unit.

Figure 6.14 Concrete storage module (PSA-Almeria) before installation of insulation. Source: Steinmann [31]/with permission of Elsevier.

Figure 6.15 Concrete storage module (400 kWh_{th}) connected to test rig, before installation of insulation.

There has been no deterioration after more than 400 cycles. A storage unit meant to operate with superheated pressurized steam (100 bar) at maximum temperatures of 500°C has also been designed using the same basic principles. According to Bahl et al. [32], the estimated specific investment costs for a storage unit with a capacity of 1100 MWh_{th} are approximately 44 c/kWh_{th}.

At the moment, options to lower the investment costs are the main focus of the ongoing development. The heat exchanger built into the storage volume accounts for the majority of these expenses. Numerous strategies to improve the storage material's effective heat conductivity have been looked into. There hasn't been much evidence of cost savings from the uniform addition of materials with high thermal conductivity. Another method lowers the number of tubes required for the heat exchanger by integrating extra heat transfer structures into the storage material.

6.12.1 Packed Bed

There isn't a wall between the HTF and the storage medium in a direct contact storage system. Storage material particles are crammed into a container, and the HTF flows through the particles. Large volume-specific heat transfer areas are possible with direct contact heat transfer. Especially for gaseous HTFs, a large effective flow cross section can minimize pressure losses. HTF and storage material need to be compatible and operate at the same pressure. An integrated packed bed storage system is part of the Jülich Solar Power Tower. The heat transfer medium in this experimental central receiver plant is atmospheric pressured air. With a storage capacity of nearly 9 MWth, the storage is cycled between 120 and 680°C.

Liquid hydrogen fuel cells can also be used to power direct contact storage systems. Here, replacing pricey liquid storage media with more affordable solids is the primary goal. Using a combination of thermal oil, sand, and gravel kept in a single tank, this method was selected at the Solar One central receiver plant. Buoyant forces keep the cold fluid at the bottom and the hot fluid at the top apart. Hot thermal oil is fed into the tank's top to charge the thermocline system. Hot oil is drawn from the top of the tank, pumped through a heat exchanger, and then returned to the bottom of the tank during the discharge process. The system reached a maximum temperature of approximately 415°C due to the thermal oil that was chosen. Melted salt was used as the HTF in an investigation into the thermocline concept.

6.12.2 Solid Particles

For solar chemistry applications requiring high temperatures, receivers with direct absorption of concentrated solar radiation in solid particles are thought to be an appealing option. In essence, this method can also be selected as an energy storage device. Although it is an appealing option, there are a number of issues that need to be resolved before commercial-scale CSP applications can be realized. One of these is the direct absorption of solar energy in nonpressurized solid storage materials. It is necessary to guarantee the particles' long-term stability and to take into account the parasitic load required for particle transportation. Heat transfer from the particles to the power cycle's working fluid is complicated, and the piping used to transport the particles will be subjected to heavy mechanical loads at high temperatures.

6.13 Latent Heat Storage Concepts

To store energy, latent heat storage systems use the enthalpy change of a substance going through a phase change, typically from solid to liquid. The ability to store energy within a limited temperature range near the phase change temperature is the main benefit of storage concepts that use the latent heat associated with the change of state of the storage material. Interest in latent heat storage systems has increased in CSP technology due to the development of absorbers that directly generate steam. In this case, it is typically not cost-effective to apply storage concepts utilizing sensible heat storage. This is illustrated in Figure 6.16. A sensible heat store must be charged and discharged using a steam process, which means the charging steam system must operate at a much higher saturation temperature.

When compared to the charging steam, the steam with recovered energy will have a significantly lower potential for power generation (exergy). This significant energy loss can be prevented if the latent heat of vaporization for the steam is captured and released by a suitably matched latent heat-based energy store, and if the superheat component is satisfied by a sensible heat store. The phase change material (PCM) selection is the first step in the development of a latent heat storage system. The phase change temperature should be appropriate for the given application. This typically necessitates that the storage material's melting point be relatively close to the saturation temperature that results from the intended steam turbine operating pressure. There should be no change in the melting temperature or melting enthalpy over a number of cycles for the phase change process to be both physically and chemically reversible. The melting temperature range for candidate PCMs is 120–440°C for CSP and solar process heat application using direct steam generation in the absorbers. Materials with melting temperatures in this temperature range are listed in Table 6.3. Thermal conductivity and the specific heat of fusion are two additional pertinent physical requirements for materials intended for storage.

Because of their high cost, some metals, like lead or tin, cannot be classified as PCMs. The low thermal conductivity of the remaining candidate materials is one of their characteristics. Therefore, in order to overcome the constraints brought about by the low heat conductivity of the storage material, the development of PCM storage systems necessitates the identification of cost-effective heat transfer concepts. The method used to guarantee adequate heat transfer between the storage material and the HTF allows the different PCM storage concepts

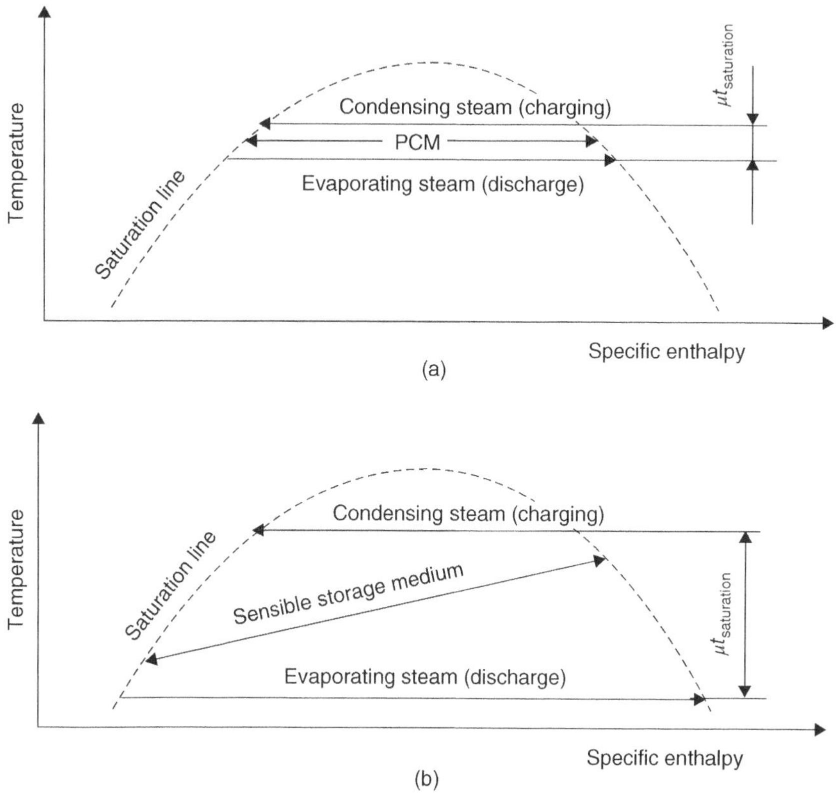

Figure 6.16 Necessary reduction of saturation temperature for a system using steam as working fluid. Comparison of latent heat storage concept (a) and sensible heat storage concept (b).

Table 6.3 Examples for PCM with melting temperatures in the temperature range relevant for CSP.

PCM	Melting temperature (°C)	Density (kg/m^4)	Thermal conductivity (W/mK)	Heat of fusion (kJ/kg)	Volume-specific latent heat (kWh$_{th}$/m^4)	Capacity-specific media costs (c/kWh$_{th}$)
KNO_4-$NaNO_2$-$NaNO_4$(eu)	142	2,000	0.5	60	44	65
KNO_4-$NaNO_4$(eu)	222	2,000	0.5	100	55	28
Tin	242	7,200	60	59	118	600
$LiNO_4$	252	1,800	0.6	460	180	60
$NaNO_4$	406	1,900	0.5	175	96	14
Lead	427	11,400	45	24	72	100

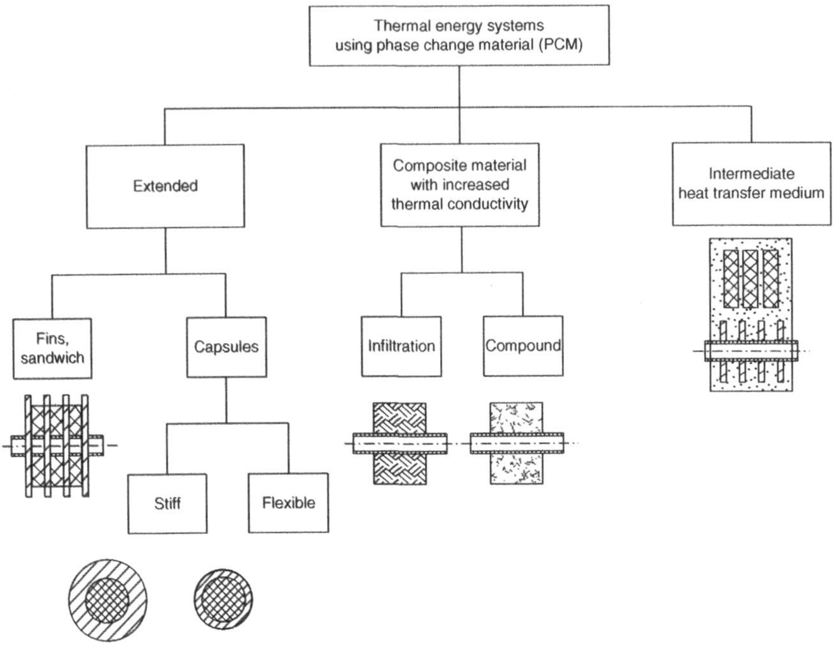

Figure 6.17 Concepts for latent heat energy storage.

to be identified (Figure 6.17). These ideas entail boosting the PCM volume with additional materials. To determine the acceptable fraction for additional materials, the volume-specific costs of these materials should be compared to the volume-specific costs of PCM (roughly 1600 c/m^4). Since the storage material frequently experiences a large volume change during the phase change, PCM storage concepts are typically more complicated than sensible heat storage systems. The ensuing sections address each of these ideas in turn.

6.14 Phase Change Material (PCM) Concept with Extended Heat Transfer Area

With this method, the average heat transfer distance within the PCM is constrained. The effective surface of the tube is increased when finned tubes are used. The idea is to use less-expensive nonpressurized thermally conductive structures in place of pricey pressure pipes. Finned tubes utilized in PCM storage are not the same as those used in heat exchangers in a number of ways.

The fins' material must be corrosion resistant in a PCM environment (such as molten salt), and they must be able to withstand thermomechanical stress brought

Table 6.4 Materials for extended surface heat transfer.

Material	Thermal conductivity (W/mK)	Density (kg/m^4)	Volume-specific material costs (c/m^4)
Aluminum	200	2,700	7,000
Graphite foil	150	1,000	9,000
Carbon steel	40	7,800	14,000
Stainless steel	20	7,800	19,000
Copper	450	8,800	45,000

on by the PCM's volume variation within a narrow temperature range during phase changes. The distance between parallel tubes is also greater than it is in conventional heat exchangers. Particular attention should be paid to the material used for the fins (Table 6.4). Steel is a strong material for pressurized tubes, but because of its high volume-specific costs and moderate thermal conductivity, it is not the best option for the fins. As an alternative, fins composed of graphite or aluminum are possible. Heat exchanger for PCM storage featuring high aluminum fins exhibit relatively high thermal conductivity and less-expensive volumetric costs. If nitrates are used as PCM, the temperature range for these fin materials is restricted due to corrosion resistance. However, in terms of the range of the saturation temperature for water, aluminum can cover the relevant temperature range in full. Several DLR research projects have shown the feasibility of the finned tube concept embedded in PCM (sandwich concept) using either graphite or aluminum fins. Aluminum fins have been used in the design of a 700 kWh_{th} PCM storage unit intended to operate at 100 bar of steam. In Carboneras, Spain, this storage unit was incorporated into a test loop that was connected to the Litoral of Endesa power plant.

An additional method for expanding the heat transfer area is the macroencapsulation of PCM. In this instance, PCM-filled capsules are piled inside a pressure vessel. For cooling and low-temperature applications, macroencapsulated low-temperature PCMs are commercially available. Capsules with metallic walls are required for CSP applications (Figure 6.14). PCM containers are essentially classified as thick-walled (stiff) with varying pressures or thin-walled (flexible) with equal pressure inside and outside. In systems that use molten salts, the walls need to be at least that thick in order to guarantee a long enough lifespan in terms of corrosion. Thus, for nitrates as the PCM, a design utilizing flexible containers is not feasible. The requirement to include a gas volume to account for the PCM's expansion during melting is a major disadvantage of stiff capsules (Steinmann and Tamme 2008). PCM capsules can occupy about 60% of the pressure vessel's volume. DLR designed and manufactured a test unit for use in a laboratory setting. A pressure vessel was used to store cylindrical capsules with a total mass of 4.5 kg of eutectic $NaNO_4$-KNO_4 (Figure 6.14).

6.15 Conclusion

CPV systems offer an innovative and efficient approach to solar energy generation by utilizing optical devices, such as lenses or mirrors, to focus sunlight onto high-efficiency PV cells. This approach allows for higher energy conversion efficiency compared to conventional flat-plate solar systems, particularly in regions with high direct sunlight. CPV technology benefits from reduced material use in PV cells, as smaller, high-performance cells can be used in combination with concentrators to achieve greater energy output.

CPV systems are particularly suited for deployment in areas with high solar irradiance, where they can generate significant power while using less land compared to traditional PV systems. The integration of dual-axis tracking further enhances their ability to capture maximum sunlight throughout the day.

Despite these advantages, CPV systems face challenges in terms of high initial costs, complex system designs, and sensitivity to diffused light, which limits their effectiveness in regions with lower direct sunlight or frequent cloud cover. Advances in materials, cost-effective manufacturing, and system optimization are essential to overcoming these barriers.

In conclusion, while CPV systems present significant potential for achieving high-efficiency solar power generation, their widespread adoption will depend on continued technological innovations and cost reductions. With further development, CPV could play a critical role in the global transition to renewable energy, especially in sunny regions where their efficiency advantages can be fully realized.

References

1 Zou, Y., Qin, C., Liu, H. et al. (2024). Concentrating photovoltaic systems: a review of temperature effects and components. *Journal of Thermal Analysis and Calorimetry* 149 (4): 1301–1329.
2 Abd-Elhady, M.M., Elhendawy, M.A., Abd-Elmajeed, M.S., and Rizk, R.B. (2025). Enhancing photovoltaic systems: a comprehensive review of cooling, concentration, spectral splitting, and tracking techniques. *Next Energy* 6: 100185.
3 Shanks, K. (2024). Concentrator and multijunction solar cells. In: *Photovoltaics Beyond Silicon*, 499–522. Elsevier.

4 Welford, W. and Winston, R. (1989). *High Collection Nonimaging Optics*. Academic Press.

5 Cancro, C., Palladino, V., Borriello, A. et al. (2024). Design of a novel hybrid concentrated photovoltaic–thermal system equipped with energy storages, optimized for use in residential contexts. In: *Solar*, vol. 4, 526–554. MDPI.

6 Said, Z., Bellos, E., Ali, H.M. et al. (2025). Nanofluids, turbulators, and novel working fluids for heat transfer processes and energy applications: current status and prospective. *Applied Thermal Engineering* 258: 124478.

7 Kowsar, A., Debnath, S.C., Shafayet-Ul-Islam, M. et al. (2024). Progress in multijunction solar cells. *arXiv preprint arXiv:2405.01550*.

8 Green, M. (1998). *Solar Cells. Operating Principles, Technology and System Applications*. Prentice-Hall.

9 Abdulla, H., Sleptchenko, A., and Nayfeh, A. (2024). Photovoltaic systems operation and maintenance: a review and future directions. *Renewable and Sustainable Energy Reviews* 195: 114342.

10 Ibrahim, T., Abou Akrouch, M., Hachem, F. et al. (2024). Cooling techniques for enhanced efficiency of photovoltaic panels—comparative analysis with environmental and economic insights. *Energies* 17 (3): 713.

11 Kurtz, S. and Geisz, J. (2010). Multijunction solar cells for conversion of concentrated sunlight to electricity. *Optics Express* 18 (S1): A73–A78.

12 Variava, J.M., Ratnadhariya, J.K., Siddiqui, M.I.H., and Sadasivuni, K.K. (2024). 3D numerical model of a concentrated photovoltaic thermal (CPV/T) system for thermal and electrical performance optimization. *Case Studies in Thermal Engineering* 61: 104823.

13 Baxter, R. (2006). *Energy Storage: A Nontechnical Guide*. PenWell.

14 Mansour, M.S., Halawa, M., Yahya, H. et al. (2024). Enhancement of PV and concentered PV panels using evaporative cooling during summer and winter seasons: a case study in Egypt. *Case Studies in Thermal Engineering* 63: 105354.

15 Reich-Weiser, C., Dornfield, D.A., and Horne, S. (2008). Environmental assessment and metrics for solar: case study of SolFocus concentrator systems. IEEE PV Specialists Conference, San Diego.

16 Thomalla, E., Köpke, P., Müller, H., and Quenzel, H. (1983). Circumsolar radiation calculated for various atmospheric conditions. *Solar Energy* 30 (6): 575–587.

17 Lorenzo, E. (2002). Energy collected and delivered by PV modules. In: *Handbook of Photovoltaic Science and Engineering* (ed. A. Luque and S. Hegedus). Wiley.

18 Ferrer-Rodriguez, J.P., Fernández, E.F., Baig, H. et al. (2018). Development, indoor characterisation and comparison to optical modelling of four

Fresnel-based high-CPV units equipped with refractive secondary optics. *Solar Energy Materials and Solar Cells* 186: 273–283.

19 Liu, Y., Zhang, Y., Zhao, Y. et al. (2021). Performance comparison of solar single-axis tracking low-concentrating photovoltaic/thermal system with different axes. *IEEE Access* 9: 28632–28645.

20 Jo, J.H., Waszak, R., and Shawgo, M. (2014). Feasibility of concentrated photovoltaic systems (CPV) in various united states geographic locations. *Energy Technology & Policy* 1 (1): 84–90.

21 Baigorri, J., Zaversky, F., and Astrain, D. (2023). Massive grid-scale energy storage for next-generation concentrated solar power: a review of the potential emerging concepts. *Renewable and Sustainable Energy Reviews* 185: 113633.

22 Xiao, M., Tang, L., Zhang, X. et al. (2018). A review on recent development of cooling technologies for concentrated photovoltaics (CPV) systems. *Energies* 11 (12): 3416.

23 Charabi, Y. and Gastli, A. (2013). Integration of temperature and dust effects in siting large PV power plant in hot arid area. *Renewable Energy* 57: 635–644.

24 Maleki, A., Ameri, M., and Pourfayaz, F. (2014). Techno-economic analysis and optimal design of an off-grid hybrid PV/wind/diesel system with battery storage. *International Journal of Engineering and Applied Sciences* 6 (4): 23–39.

25 Tina, G.M., Ventura, C., Ferlito, S., and De Vito, S. (2021). A state-of-art-review on machine-learning based methods for PV. *Applied Sciences* 11 (16): 7550.

26 Short, W., Packey, D.J., and Holt, T. (1995). *A Manual for the Economic Evaluation of Energy Efficiency and Renewable Energy Technologies*, NREL/TP-462-5173. Golden, CO: National Renewable Energy Laboratory.

27 Melia, J., Horne, S., and Klaren, A. (2010). Rational ratings. Proceedings of the 25th European Photovoltaic Solar Energy Conference.

28 Swanson, R. (2002). Photovoltaic concentrators. In: *Handbook of Photovoltaic Science and Engineering* (ed. A. Luque and S. Hegedus). Wiley.

29 Sioshansi, R. and Denholm, P. (2010). The value of concentrating solar power and thermal energy storage. Technical report NREL-TP-6A2-45833.

30 Gilli, P.V. and Beckmann, G. (1976). Design and economy of solar plants with integrated thermal energy storage. Solar Energy, Proceedings of the UNESCO/WMO Symposium, Geneva.

31 Steinmann, W-D. (2021). Thermal energy storage systems for concentrating solar power plants. In: *Concentrating Solar Power Technology*. (ed. K. Lovegrove and W. Stein), 399–440, Woodhead Publishing.

32 Bahl, C., Laing, D., Hempel, M., and Stückle, A. (2009). Concrete thermal energy storage for solar thermal power plants and industrial process heat. SolarPaces Conference 2009, Berlin.

7

Hybridization of Concentrating Solar Power (CSP) with Fossil Fuel Power Plants

7.1 Introduction

Many cutting-edge technologies are used in the operation of concentrated solar thermal power (CSP) plants, but concerns still surround the installation of systems larger than 100 MW and the potential for cost reduction to enable mass adoption.

Various technologies are being explored as ways to lower costs and increase the solar-to-electric efficiency of solar thermal power systems, but the majority of these technologies face theoretical and practical constraints. The thermal cycle has a high conversion efficiency when solar thermal energy is collected and used at a higher temperature; in contrast, solar collectors have a lower efficiency and are more expensive to produce. However, the efficiency of the thermal cycle is greatly diminished when solar thermal energy is used at a lower temperature, making solar concentrators less expensive to manufacture.

Dispatching is a major problem for thermal power plants that rely solely on solar energy. Energy storage of some kind can help offset the sporadic nature of solar energy sources. In Chapter 11, the latest developments in the field of energy storage systems are covered in detail. Integrating CSP into a traditional fossil plant is another workable solution to the dispatchability issue. The potential for hybridization to enable the economically viable utilization of solar energy at a level appropriate for energy needs is enormous.

There are two types of hybridized power plants: those that use thermochemical integration and those that use thermal integration. Hybridization in thermal integration processes combines the combustion of fossil fuels with solar radiation to heat water, saturated, or superheated steam. Fossil fuels are utilized as chemical reactants in thermochemical hybridization, and solar energy provides the process heat needed to upgrade or decarbonize the fuel to create a cleaner fuel. While thermochemical hybridization is still in its experimental and demonstration stages, thermal hybridization is already being used in industrial settings.

Concentrated Solar Power Systems, First Edition. Bellamkonda Pragathi and D. P. Kothari.

A solar hybrid power plant, as opposed to a solar-only thermal power plant, can make use of the conventional power station's existing infrastructure, which lowers the equipment investment and the cost of producing electricity. In addition, it makes it possible to avoid the issue caused by solar energy's sporadic nature. Furthermore, integrating solar energy into already-operating fossil fuel plants helps mitigate the scarcity of fossil fuels and lower greenhouse gas emissions, particularly CO_2. Therefore, creating hybridized solar and fossil fuel power plants is a workable way to hasten the wider adoption of solar thermal power technology in the short and medium terms.

This chapter discusses the different integration strategies and types of solar/fossil fuel hybridization that involve coal and natural gas. Future promising developments in solar hybridization technology are looked forward to, along with a discussion of equipment, integration system design, and hybridization technology [1, 2].

7.2 Solar Hybridization Approaches

There are several fundamental strategies for hybridizing solar and fossil fuels. This chapter's subsequent sections cover a number of these that are already in use in commercial plants as well as some cutting-edge ideas that are still being researched and developed.

Fossil backup and boosting of solar thermal plants: The most prevalent type of hybridization is the use of fossil fuel as a backup energy source by many solar thermal power plants when sunlight isn't present. With a combined capacity of 454 MW, the nine commercial solar electric generating systems (SEGS) represent the most advanced and prosperous type of solar-hybrid pure Rankine cycles. These systems employ heat exchangers to transmit thermal energy to Rankine cycle turbines via parabolic trough solar collectors and synthetic oil in a collector loop. Gas-fired backup When the steam temperature is lower than what the steam turbines require, boilers are used.

Solar-aided coal-fired power plants: Australia and China, two nations with abundant coal reserves, are good candidates for solar hybridization combined with coal-fired steam generation. In coal-fired power plants, solar heat can be easily incorporated to operate in parallel with the boiler or feedwater heaters. For instance, the extraction of steam from the turbine to heat the feedwater can be substituted with solar heat. With a capacity of less than 400 MW, this technology may be an especially appealing choice for repowering older coal-fired power plants. This system's ability to increase a coal-fired power plant's output without requiring an oversized steam turbine is one of its main advantages.

Integrated solar combined cycle (ISCC) plants: In conventional combined cycle power plants, the fossil fuel is burned in series with a steam-based Rankine cycle using a gas turbine (Brayton) cycle. Out of all the frequently utilized fossil-fired power generation technologies, this type of fossil fuel power plant offers the highest conversion efficiency.

Concentrated solar heat is added to or substituted for the exhaust gas from the gas turbine in an ISCC plant's gas-fired "combined cycle" power plant to create saturated or superheated steam. The goal of ISCC systems is to maximize the efficiency of both the pure solar input and the fossil fuel input by incorporating solar steam into the steam cycle of these plants.

Advanced systems: Unlike the systems mentioned above, which combine solar energy with the traditional Rankine cycle, advanced hybridization systems integrate solar heat and Brayton cycles to convert solar energy into electricity. Advanced hybrid power plants fall into two primary categories. In the first, solar heat is used to warm the gas turbine cycle's compressor discharge air. In the second, however, solar heat is combined with the process of decarbonizing fossil fuels through gasification, steam reforming, or decomposition to create a cleaner fuel that is referred to as "solar fuel." This chapter does not address the hybridization of solar thermochemical sources with fossil fuels. In Chapter 20, the process of producing solar fuels is covered in detail [3].

7.3 The Role of Different Concentrators

Solar collectors can be added to a traditional fossil fuel power plant to create a hybrid power system. This only contributes a small portion of the solar energy required for a sizable fossil fuel-fired plant operating independently. One important factor is the cost-effective design of the solar concentrator field, which is directly tied to finding a good balance between solar heat, the concentrator's concentration ratio, and the power plant's needs. A parabolic trough or linear Fresnel is typically needed for concentrated lower temperature solar heat (below 500°C), while a dish or tower collector is typically needed for concentrated higher temperature solar heat (above 500°C). These primary concentrator types are introduced in brief in the sections that follow.

Parabolic dish: A parabolic dish-shaped reflector is used in this system, and the receiver is positioned at the dish's focal point. The receiver's operating temperature ranges from 750 to 1,000°C, and its concentration ratio is roughly 1,000–4,000 [1, 2]. Although direct steam generation has been demonstrated, Stirling engines have been the focus of a significant amount of research on dish systems. Up to the highest temperatures and pressures possible in a fossil fuel-fueled plant, these systems could produce steam.

Solar tower: A single receiver is placed atop a tall tower at the center of a sizable heliostat field in a solar tower system. The receiver can use air, molten salts, liquid sodium, water, or steam as working materials. Its operating temperature is typically between 400 and 2,000°C, and its concentration ratio is typically between 150 and 1,500. Similar to dishes, tower systems can produce steam directly or through the use of a heat transfer fluid (HTF) at the highest temperatures and pressures possible in a traditional power plant.

Parabolic trough: This system uses mirror reflectors in the shape of a parabolic trough to focus sunlight onto receiver tubes positioned in the focal line of the trough. A trough receiver can operate in the temperature range of 50–500°C. For instance, molten salt used as an HTF in projects like the Archimede project in southern Italy can be heated from 290 to 550°C [4], while synthetic thermal oil can be heated to about 400°C [4]. The development of trough systems for direct steam production is also advanced. Though it can be applied at different stages of a steam cycle, the steam generated by troughs, whether directly or indirectly, won't be as hot as that produced by tower or dish systems.

Linear Fresnel reflector (LFR): Strips of mirrors rotate around separate parallel axes in an LFR system, as covered in Chapter 6, to reflect sunlight onto a stationary linear receiver [5]. It uses a line focus reflector array, which is comparable to a parabolic trough system. The concentration ratio is roughly 10–40, and the operating temperature ranges from 50 to 400°C [1]. For instance, LFR collector strings are used to heat water in the Puerto Errado Thermo-solar Power Plants in Spain from 140 to 270°C [4].

7.4 Process Integration and Design

Figures 7.1 and 7.2 depict a SEGS plant's flow sheet. Through the parabolic trough collector field, the HTF is heated to 494°C. It then passes through a series of heat exchangers to produce high-pressure (HP) superheated steam at 471°C and 100 bar. The cooled HTF is circulated back through the solar field after going through the HTF side of the solar heat exchangers. To generate power, the superheated steam is subsequently fed into a traditional reheat steam turbine. Condensed turbine exhaust steam is fed back into the heat exchangers and used in an oil heater in parallel with the solar field (SEGS VIII–IX). The data relating to the SEGS power plants is given in Table 7.1.

7.4.1 Economic Effect

The economic performance of trough plants operating in hybrid and solar-only modes is compared in Table 7.2. The cost data that is provided is out of date

Figure 7.1 SEGS III–VII plants located at Kramer Junction.
Source: A.R. Akradecki/Wikimedia Commons/CC BY SA 3.0.

and should only be utilized to determine the relationships between the different system configurations. For the 40 MWe SEGS VI trough plant in 2002, the cost of power was quoted at 0.17 $/kWh for a solar-only plant and 0.141 $/kWh for a hybrid plant with a 25% fossil backup. The utilization of ball-joint assemblies and new solar receivers is largely responsible for the anticipated one-third increase in solar-to-electric efficiency in plants scheduled for construction in the near future, compared to the original SEGS plants. Because of the increased plant capacity and the more effective solar field, unit capital costs are lower and the necessary solar field size is reduced. For a solar-only plant, the levelized cost of energy (LCOE) is lowered to 0.11 $/kWh, while for a hybrid plant, it is reduced to 0.096 $/kWh. Overall, a solar power plant's electricity costs are 12–17% cheaper when hybridized solar energy is combined with fossil fuels than when solar energy is used [5].

7.5 Hybridization Process and Arrangement

Three different arrangements are used in solar-aided coal-fired power plants.

- **Solar-aided with boiler drum**: In this hybridization setup, the boiler and solar collector field run concurrently. Saturated steam is created when the sun's heat, which is approximately 400°C, transforms the feedwater from the boiler's economizer. This steam is then redirected into the boiler's drum using fossil fuels.

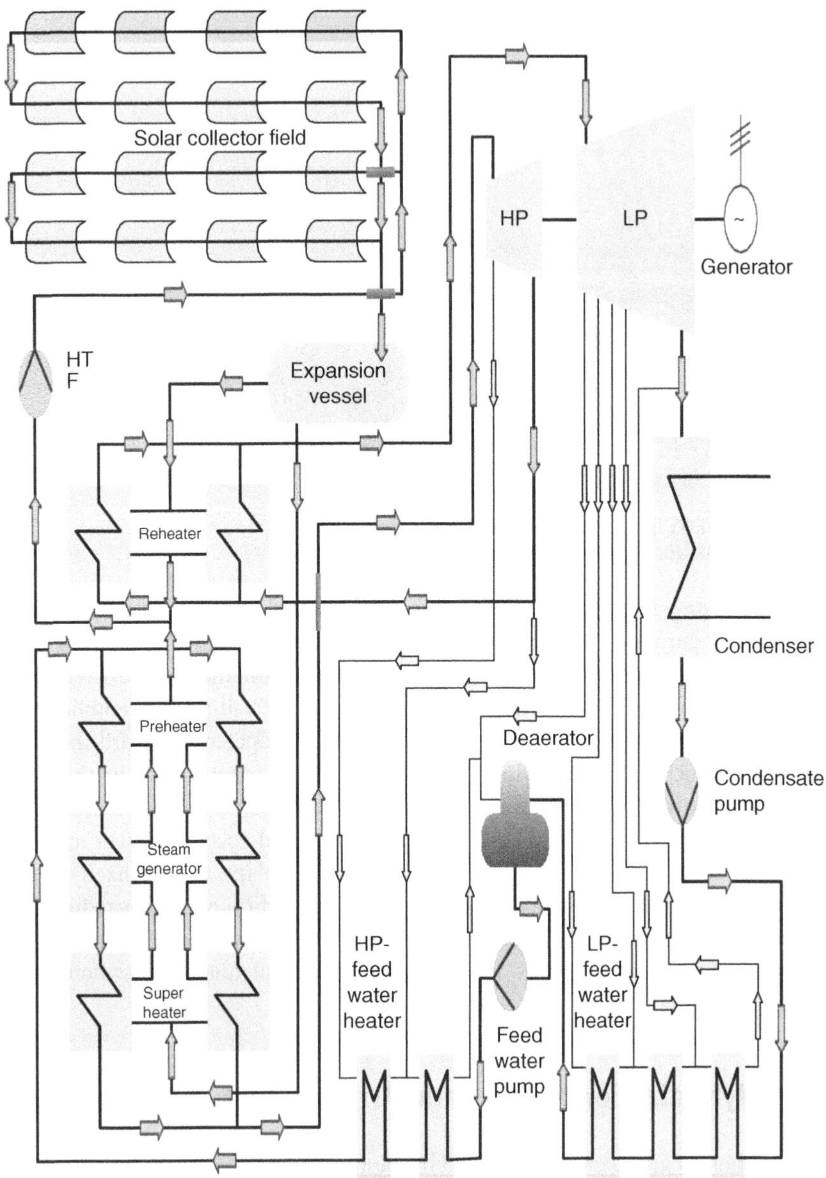

Figure 7.2 Flow diagram of the SEGS plant for pure solar mode. Source: Reproduced from Giuliano et al. [7]/with permission of Elsevier.

Table 7.1 SEGS power plant data from NREL.

Plant name	Net output (MW)	Solar field outlet (°C/bar)	Solar field area (km^2)	Solar turbine efficiency (%)	Solar field thermal efficiency (%)	Solar-to-electric efficiency	Dispatchability
SEGS IX	80	490/100 reheat	0.484	47.6	50	14.6	HTF heater
SEGS VIII	80	490/100 reheat	0.464	47.6	54	14	HTF heater
SEGS VII	40	490/100 reheat	0.188	47.5	44	12.4	Gas boiler
SEGS VI	40	490/100 reheat	0.194	47.5	44	12.4	Gas boiler
SEGS V	40	449/40	0.251	40.6	44	4.2	Gas boiler
SEGS IV	40	449/40	0.240	40.6	44	4.2	Gas boiler
SEGS III	40	449/40	0.240	40.6	44	4.2	Gas boiler
SEGS II	40	416/40	0.190	29.4	44	4.7	Gas boiler
SEGS I	14.8	407/45	0.084	41.5	45	9.4	4-hour TES

Table 7.2 Solar-only and hybrid operation comparison.

Site: Kramer Junction, 40 MW SEGS plant, Near-Term Trough Plant				
	Solar-only	**Hybrid (25%)**	**Solar-only**	**Hybrid (25%)**
Plant size, net electric (MW)	40	40	50	50
Collector aperture area (km^2)	0.188	0.188	0.412	0.412
Thermal storage (h)	0	0	0	0
Solar-to-electric efficiency (%)	4.6	4.7	14.9	14.1
Plant capacity factor	22.2	40.4	29.2	49.6
Capital cost ($/kW)	4008	4204	2745	2949
O&M cost ($/kWh)	0.046	0.044	0.024	0.018
Fuel cost ($/kWh)	0	0.014	0	0.01
LCOE (2002$/kWh)	0.17	0.141	0.11	0.096

- **Solar coupled with feedwater**: In this setup, solar radiation operating at approximately 400°C serves as a heat source to warm the feedwater rather than drawing steam from the turbine. The powerhouse feedwater heaters are connected in parallel to the solar field. The feedwater can be heated to the necessary temperature by the 400°C solar heat [6, 7] boiler, which is typically at 220°C. With the same amount of input coal, the steam turbine produces more output work than it did prior to hybridization because the extra steam that does not need to be extracted is further converted in the turbine. This allows for the updating of this traditional coal-fired power plant to a larger capacity system without resulting in significant financial outlays.
- **Solar-powered superheater**: In this setup, some of the superheated steam that is injected into the steam turbine is produced using solar heat. The boiler and solar collector field run simultaneously.

The simplest to execute of the three configurations mentioned above is the one that combines solar heat and feedwater. With the help of solar technology, coal-fired power plants that have been repowered and have a capacity less than 400 MW can now have more capacity. This solar hybridization power technology eliminates the need for solar energy storage devices, in contrast to solar-only thermal power plants. Additionally, a solar-powered feedwater heater runs concurrently with the current feedwater heaters. This ensures that even in the event of low solar radiation, the hybridized power plant will run at full capacity [1].

7.6 Case Study Design

The hybridization of solar energy with coal-fired power technology is suitable for countries rich in coal resources such as China. In northwest China, a large number of old coal-fired power plants with small capacity are still in use, which contributes to the high pollution levels in the environment (Figure 7.3). The northwest of China, particularly the Xinjiang area, has an abundance of solar radiation, averaging 5,440–6,670 MJ/m^2/year, and 2,550–4,500 hours of sunshine annually. The design flow sheet for a hybridization system that might be installed in a 145 MWe coal-fired power plant in Kashi, Xinjiang, where there is an average of 4,047 hours of sunshine per year, is shown in Figure 7.4.

The current 145 MW coal-fired power plant retains its four low-pressure (LP) heaters and its two HP heaters in this standard configuration. These run concurrently with the solar feedwater heating system [1].

The HP-extracted steam is replaced by concentrated solar heat at about 400°C to heat the feedwater to 240°C before it is injected into the boiler's economizer. The total mass flow of the feedwater and the regenerative subsystem's inlet temperature can be kept at the same values as before repowering when the direct

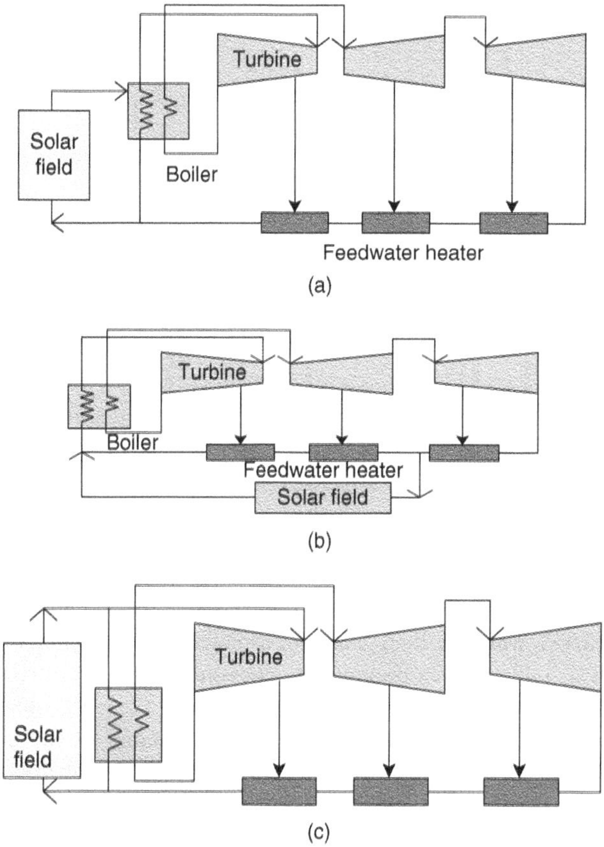

Figure 7.3 Schematics of three solar-aided coal-fired processes: (a) boiling process; (b) feedwater heating process; (c) boiling and feedwater heating process.

normal irradiation (DNI) drops. To keep the boiler economizer's inlet temperature constant, the feedwater flow in the solar collector is reduced, but the flow in the earlier feedwater heaters is increased. This allows the repowered plant to run at full capacity when there is little solar radiation.

From a design standpoint, the hybrid system's total power output raises the coal-fired power plant's capacity to 152 MW, resulting in a decrease in coal consumption from 419.75 to 284.24 g/kWh. This higher output is a result of solar energy utilization. However, when considering the conversion of solar energy into electrical power, this 17 MW increase in capacity is comparable to the capacity of a power plant that uses only solar energy. This hybrid plant can achieve an annual solar-to-electricity efficiency of 15%, which is roughly 4% points greater than the most advanced SEGS VI technology.

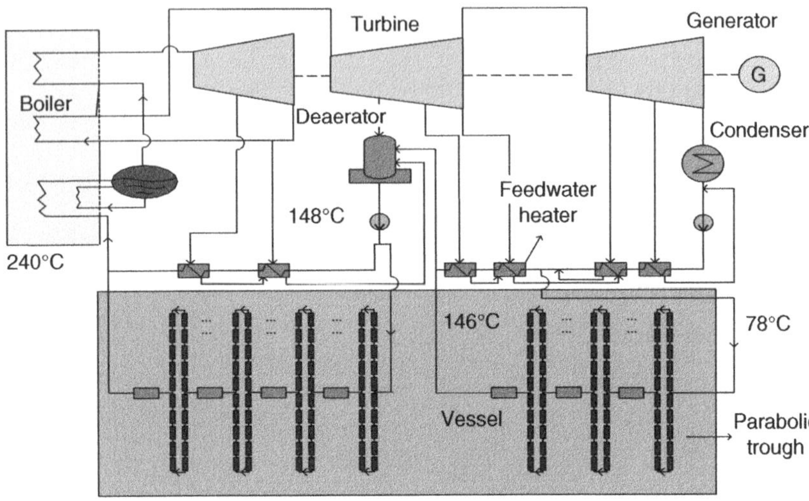

Figure 7.4 A design flow sheet for the typical 145 MW solar hybrid coal-fired power plant in Xinjiang, China.

The 40 MW SEGS plant in standard configuration is estimated to have a capital cost of nearly 4,000 $/kW, whereas the typical unit capital cost is estimated to be nearly 2,007 $/kW [9]. Table 7.3 is a preliminary assessment of the required investment for this plant [6].

It is important to remember that the concentrated solar heat range of 400°C is typically utilized to heat water instead of producing electricity. Fortunately, hybridization of a solar system with a coal-fired power plant may allow low-grade solar heat to provide electricity.

Table 7.3 Preliminary evaluation of investment.

Total investment cost	**Million $**	**44.78**
Operation and maintenance (O&M) cost	Million $	0.70
Annual average coefficient of device investment (CRF)	$	0.12
Annual cost of device investment	Million $	4.09
Net increased generation power	MWe	17.08
Increased electricity	Million kWh/yr	52
Net increased profit	Million $/yr	4.70
Payback period	Y	9.49
Solar-generation cost	$/kWh	0.09
Specific investment cost	$/kW	2,007

7.7 Potential of Systems in China

China's installed capacity of traditional coal-fired power plants is projected to reach 599 GW by the end of 2009. It was estimated that small units (less than 400 MWe) could produce about 40.57%, or 184 GW. 16.2 million tons less coal would be used annually if these conventional plants were converted to solar-aided coal-fired systems, assuming 2,500 hours of sunshine annually.

Several businesses and academic institutions in China are encouraging the commercial use of this type of solar hybrid plant. The Chinese Academy of Sciences, North China Electric Power University, and the Institute of Engineering Thermophysics, for instance, are creating the technology required for system integration. Prefeasibility studies have been started by power generation companies, including Datang, Huadian, Guodian, and others, for hybrid plants that will be constructed in Northwest China. The Chinese National Natural Science Foundation and the Ministry of Science and Technology are supporting the development of solar hybridization plants, which have been identified as a high-priority technology in the near future. They are recognized as a suitable and useful choice for Chinese solar thermal power plants.

7.7.1 Integrated Solar Combined Cycle (ISCC) Power Plants [7]

This section will examine ISCC power technology. This technology integrates solar energy into the steam cycle of a combined cycle power plant. The steam generated by solar heat may be either directly into the combined cycle's steam turbine or into various locations within the heat recovery steam generator (HRSG). Figure 7.5 displays a schematic diagram of an ISCC plant.

The primary characteristic of this method of solar hybridization is that more power is produced in the combined cycle's steam cycle without requiring the use of additional fuel; in other words, all of the extra power produced is "free" in terms of fuel. In order to accomplish this, the steam turbine's size may be increased in comparison to the unit required in a purely gas-fired application. When there is no solar energy input, the massive steam turbine operates at partial load. An ISCC plant's fuel consumption can also be decreased by using solar thermal input. When solar input is available, the gas turbine in this method operates at a partial load. It should be noted that lowering the gas turbine's fuel consumption also lowers its power and exhaust energy.

Two primary questions need to be taken into account when planning to incorporate solar-generated steam into a combined cycle: (a) How much solar energy should be integrated into the combined cycle? and (b) Where in the steam cycle is the ideal location to introduce the steam produced by solar radiation? It goes without saying that the optimal way to integrate steam produced by solar

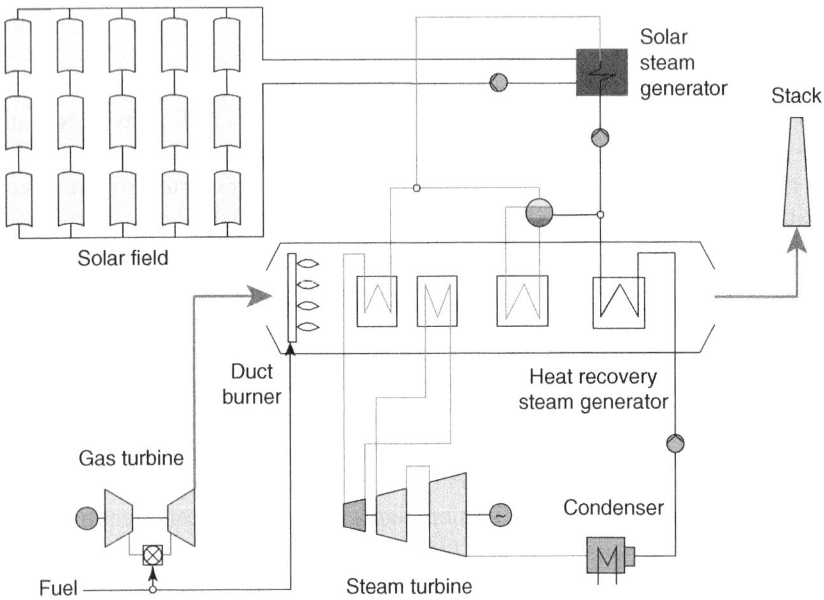

Figure 7.5 Diagram of an ISCC power plant with a single-pressure-reheat steam cycle.

technology depends greatly on the steam conditions that technology can produce. The integration of different solar technologies into combined cycle power plants is covered in the following sections.

7.8 Process Integration and Design

ISCC technology can be categorized into three types on the basis of its fluid temperature capability: high temperature: >500°C; medium temperature: 400°C; low temperature: 250–400°C. The medium temperature type has been the most widely implemented and will be discussed first.

Medium-temperature solar technology: A parabolic trough is used in the majority of medium-temperature solar technology applications. In integrated gasification combined cycle (IGCC) plants, the integration of additional HP saturated steam into an HRSG is also suggested (Figure 7.6). Up to 480°C of steam can be produced by parabolic trough systems. In IGCC plants, it's also fairly typical to send heated feedwater out of the HRSG. Keep in mind that it's crucial to supply the solar boiler with feedwater from the appropriate place in the steam cycle. The HP feedwater pump's discharge is the most practical place for this. The HTF exiting the steam generator is at 290°C, which is comparable to what is seen in SEGS plants. To maximize feedwater heating in the HRSG

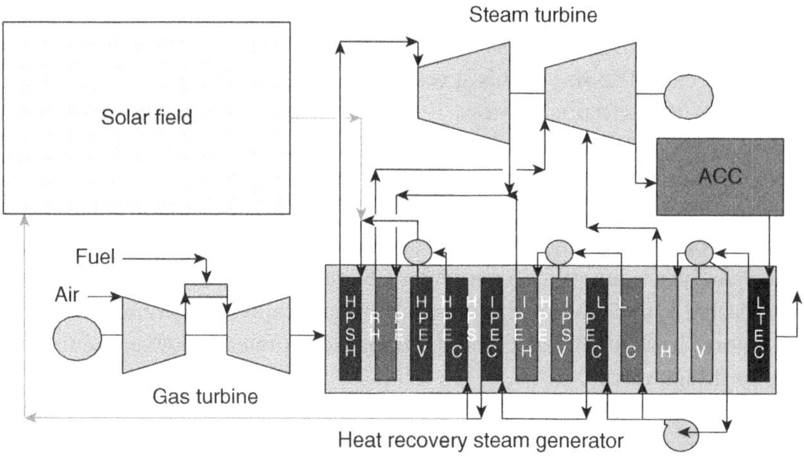

Figure 7.6 A medium-temperature solar IGCC. HPSH, high-pressure superheater; RH, reheater; HPEV, high-pressure evaporator; HPEC, high-pressure economizer; IPSH, intermediate-pressure superheater; IPEV, intermediate-pressure evaporator; IPEC, intermediate-pressure economizer; LPSH, low-pressure superheater; LPEV, low-pressure evaporator; LTEC, low-temperature economizer; ACC, air cooling condenser. Source: Reproduced from El-Katiri and Husain [8] with permission of Elsevier.

and minimize feedwater heating in the solar field, the feedwater temperature should be approximately 260°C.

High-temperature solar technology: HP, 565°C superheated steam can be produced by solar tower systems. Under these circumstances, superheated steam produced by the sun might be admitted straight into the HP steam line leading to the steam turbine. Furthermore, the power tower can reheat steam in a manner similar to that of an HRSG. This enables the solar boiler to handle the superheating and reheating of the solar steam with the least amount of influence on the HRSG.

Low-temperature solar technology: Up to 270°C/55 bar of saturated steam has been produced using the majority of linear Fresnel systems. The steam cycle's HP system cannot be integrated with this pressure. Here, you have two choices: either produce saturated steam at a pressure of about 40 bar and admit it to the cold reheat line or produce steam at a pressure of about 5 bar and admit it to the LP steam admission line [8, 9].

7.9 Major Equipment Design

The design mainly involves the integration of the solar field input with the combined cycle equipment, particularly the HRSG, steam turbine, and heat sink.

Heat recovery steam generator (HRSG): It is evident that the majority of the time, one of the HRSG sections needs to have the steam produced by the solar field reinjected. The size of this specific HRSG section would be a significant obstacle in the selection process. It is possible to accommodate both fossil and solar-generated steam; alternatively, the section could be sized to only accommodate the fossil portion.

Steam turbine: Sizing the steam turbine, which is related to both steam mass flow and pressure, is one of the key choices in determining how much solar steam can be produced (for a particular CSP technology). Cycle efficiency will be impacted by variations in these two factors, especially in nighttime or evening operations. It is recommended to maintain the volumetric flow to the turbine near its design value.

Balance of plant (BOP): Pumps, motor control centers (MCCs), pipes, and cabling are among the BOP components that are impacted by the solar-generated steam contribution.

The electrical generator needs to be able to handle the extra capacity if the goal of the solar-generated steam is to increase the plant's power output. Due to a lack of water, air-cooled condensers (ACCs) rather than traditional wet cooling towers are more frequently used as the heat sink (condenser cooling) in these plants. In any scenario, the heat rejection load is adjusted or increased to account for the specific environmental factors.

7.10 Typical Demonstration Plant and Project

One of the first ISCC plants to be built was the Yazd Solar Thermal Power Plant in Iran, which became operational in 2009. The plant comprises two gas turbine units with nearly 150 MW capacity each, one steam unit with nearly 150 MW capacity, together with a solar thermal unit with 17 MW equivalent capacity. With a total capacity of 467 MW, the Yazd power plant was the eighth largest solar power plant in the world at the start of 204.

In Ain Beni Mathar, Morocco, an ISCC project of 472 MW is being built, supported by Global Environmental Facility, as shown in Figure 7.7. The plant includes a parabolic trough solar component of 20 MW (180,000 m^2) with an expected annual net production of 4.548 GWh. The solar output is estimated at 1.14% of the annual production, representing 40 GWh per year. The plant started electricity production in May 2011.

Abener has recently constructed a second ISCC power plant in Hassi'Mel, Algeria. The complex comprises a 140 MW combined cycle, with a gas turbine power in the order of 80 MW and a 75 MW steam turbine. A 25 MW solar field,

Figure 7.7 IGCC plant in Ain Beni Mathar, Moroc. Source: With permission from Elsevier.

requiring a surface of around 180,000 m^2 of parabolic the source of non-fossil energy is mirrors. The project is the first privately financed solar thermal plant in North Africa, with investments totaling close to $140 million. In July 2011, the plant began producing electricity.

Currently under construction in Egypt is a 140 MW total capacity plant. With a $50 million grant from Global Environmental Facility, it boasts a substantial 40 MW solar contribution. Production is scheduled to begin on December 40, 2012.

An existing 700 MW power plant in Italy is getting a 40 MW solar field added to it. Three more ISCC plants are slated to be built in Florida and California, and one is currently being constructed in Victorville, California. Global Environmental Facility approved a 150 MW ISCC plant in Mexico in 2006, and a 40 MW solar contribution is planned for a 150 MW ISCC plant in India.

At the Martin Next Generation Solar Energy Center, US utility company Florida Power and Light (FPL) inaugurated the nation's first hybrid solar thermal facility in 2010. It was connected to an already-existing combined cycle power plant. The largest of FPL's solar facilities, the 75 MW CSP installation spans about 500 acres and comprises about 180,000 mirrors at the current plant location, producing up to 2.8 GW at present. ISCC projects that are presently in an advanced development stage are listed in Table 7.4.

One practical problem with ISCC systems is that they force the steam turbines to operate at reduced efficiency when solar energy is unavailable. Put differently, only about 2,000 of the combined cycle's 6,000–8,000 operating hours are spent using solar steam. As a result, ISCC systems are only thought to have short-term prospects because the achievable solar share is less than 10%.

Table 7.4 ISCC projects under development.

Country	Technology	Capacity (MWe)	Solar capacity (MWe)	Solar share (%)	DNI (kWh/m^2/yr)	Construction status on December 2011
Iran, Yazd	ISCCS PT	467	17	4.6	2,500	Complete
Algeria, Hassi R'mel	ISCCS PT	150	25	16.7	2,400	Well advanced
Morocco, Ain Beni Mathar	ISCCS PT	472	20	4.2	2,400	Well advanced
Egypt, Kuraymat	ISCCS PT	140	40	28.6	2,400	Well advanced
US, Victorville, CA	ISCCS PT	564	50	8.9	2,200–2,600	Well advanced
US, Indiantown, FL	ISCCS PT	1,125	75	6.7	–	Well advanced
Italy, Siracusa	ISCCS PT	740	40	4.1	2,100	Well advanced
US, Palmdale, CA	ISCCS PT	570	50	8.8	2,200–2,600	Well advanced
Mexico, Sonora State	ISCCS PT	500	40	6.0	2,600	Well advanced

7.10.1 Advanced Hybridization Systems

Advanced hybridization refers to the combination of solar energy with a gas turbine cycle. There are two main categories: systems that use solar heat to preheat the compressor discharge air in a gas turbine cycle and those that use solar heat to decarbonize fossil fuel for electricity generation. The latter also involves high-temperature and mid-temperature thermochemical hybridization processes. Here, we focus on the solar preheating of air and mid-temperature solar thermochemical hybridization technology.

7.11 High-Temperature Solar Air Preheating

The concentrated solar power can be used to heat pressurized air in a gas turbine before it enters the combustion chamber. There is thus no need to use fossil fuels to preheat the air from 400 to 1200°C, and the solar heat is converted with all the high thermal efficiency of a modern recuperated or combined gas turbine cycle.

Solar air preheating has great potential to reduce the costs of solar thermal power. In addition, the concept could be applied to a wide range of power levels (from 1 to 100 MW). Projects involving solar air preheating are discussed below.

Typical projects: The SOLGATE project investigated the concept of combining solar energy with fossil fuel in a hybrid plant and showed its feasibility by proving that a gas turbine could be modified to allow dual operation. Tests were carried out at different operating temperatures, which resulted in different solar shares and varying electricity generation costs. In the EU-funded SOLGATE project demonstrated at the Plataforma Solar de Almería (PSA), pressurized air temperatures of up to 960°C were achieved at the receiver exit. The aim of the closely related REFOS project is to develop, build, and test modular pressurized volumetric receivers under operating conditions representative of those encountered when receivers are integrated with gas turbines. The main focus of the project is the testing of solar air preheating, accompanied by basic research into materials. The project is led by DLR and carried out in cooperation with CIEMAT, Spain, and Germany.

7.12 Solar Thermochemical Hybridization Plant

Chapter 20 addresses solar fuels in detail. Any solar-derived fuel can be used in a combined cycle power plant with a suitably modified combustion system. One of the most obvious strategies in the short term is solar-driven natural gas reforming. This could be carried out together with combined cycle power generation, which would otherwise burn natural gas directly.

7.12.1 Case Study of Medium Temperature Thermochemical Hybridization

This section briefly discusses one example of a cost-effective mid-temperature solar thermochemical hybridization plant. Figure 7.8 shows the arrangement of the mid-temperature solar thermochemical hybridization combined cycle using methanol decomposition proposed by the group in which the authors of this chapter currently work, at the Institute of Engineering Thermophysics, Chinese Academy of Sciences. It is composed of two main integrated processes: a solar thermochemical process at around 200–400°C and a gas turbine combined cycle with dual-pressure heat recovery steam generation. The solar thermal heat collected at 200–400°C supplies the heat required for methanol decomposition. The solar fuel produced with syngas (CO and H_2) acts as a fuel to drive the combined cycle, which then produces electricity. The net solar-to-electric efficiency is expected to be about 45%, which is related to the energy-level upgrade of mid-temperature solar heat to a high level of chemical energy in the fuel.

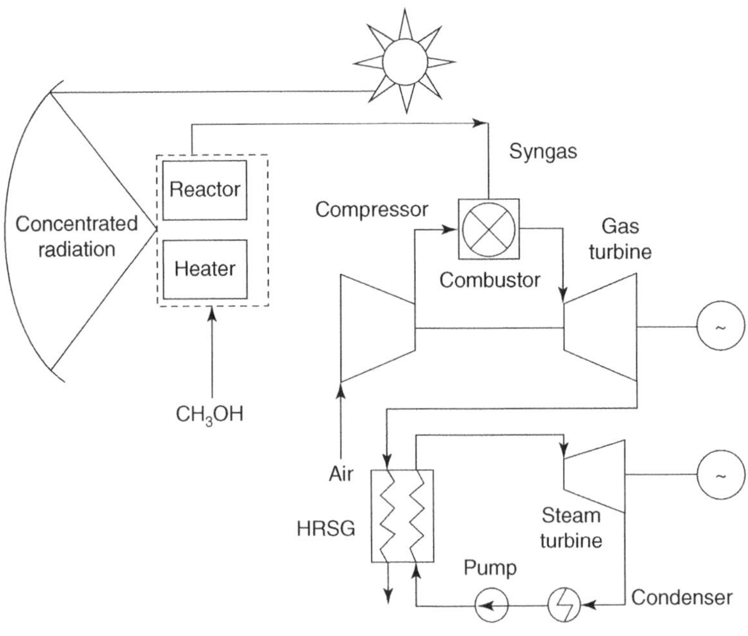

Figure 7.8 Schematic of the new solar/methanol combined cycle hybrid plant.

7.13 Conclusion

Hybridization of CSP systems with fossil fuel power plants presents a practical and transitional solution for enhancing energy efficiency while reducing carbon emissions. By combining the reliability of fossil fuel plants with the renewable energy input from CSP, this hybrid approach addresses the intermittency of solar power and ensures a steady and stable energy supply, even when solar irradiance is low. Such systems can reduce fossil fuel consumption by using solar energy to preheat working fluids, generate steam, or contribute directly to electricity production.

The integration of CSP with fossil fuel plants can lead to improved fuel efficiency, lower greenhouse gas emissions, and more flexible power generation, especially in regions with high solar potential. Furthermore, it provides an opportunity to extend the lifespan of existing fossil fuel plants while transitioning to cleaner energy sources.

However, the success of hybrid CSP-fossil fuel plants depends on overcoming technical challenges, such as optimizing system integration, managing operational complexities, and ensuring cost-effectiveness. Advances in thermal energy storage, control systems, and operational strategies will be key to maximizing the benefits of hybrid systems.

References

1 Khaled, M.M., Hossain, T., Navid, W.U. et al. (2024). Thermal analysis and dynamic attributes of a sustainable CSP-fossil hybrid power plant utilizing organic Rankine cycles for enhanced plant performance. *International Journal of Thermofluids* 22: 100696.

2 Rech, S., Lazzaretto, A., and Grigolon, E. (2018). Optimum integration of concentrating solar technologies in a real coal-fired power plant for fuel saving. *Energy Conversion and Management* 178: 299–310.

3 Akroot, A., Almaktar, M., and Alasali, F. (2024). The integration of renewable energy into a fossil fuel power generation system in oil-producing countries: a case study of an integrated solar combined cycle at the sarir power plant. *Sustainability* 16 (11): 4820.

4 Khan, M.I., Gutiérrez-Alvarez, R., Asfand, F. et al. (2024). The economics of concentrating solar power (CSP): assessing cost competitiveness and deployment potential. *Renewable and Sustainable Energy Reviews* 200: 114551.

5 Pramanik, S. and Ravikrishna, R.V. (2017). A review of concentrated solar power hybrid technologies. *Applied Thermal Engineering* 127: 602–637.

6 Li, X., Li, T., Liu, L. et al. (2023). Operation optimization for integrated energy system based on hybrid CSP-CHP considering power-to-gas technology and carbon capture system. *Journal of Cleaner Production* 391: 136119.

7 Giuliano, S., Buck, R., and Eguiguren, S. (2011). Analysis of solar-thermal power plants with thermal energy storage and solar-hybrid operation strategy. *Journal of Solar Energy Engineering* 133 (3): 031007.

8 Peterseim, J.H.M. (2014). Enabling concentrating solar power in Australia: an investigation of the benefits and potential role of concentrating solar power and non-conventional fuel hybrid plants in Australia's transition to a low-carbon energy future. Doctoral dissertation.

9 Kurup, P., Kwasnik, T., Roberts, B.J., and Wendelin, T. (2019). *Initial Thermal Energy Yield Potential for the Use of Concentrating Solar Power (CSP) for Coal Hybridization in India* (No. NREL/TP-6A20-74024). Golden, CO (United States): National Renewable Energy Laboratory (NREL).

8

Grid Integration of PV Systems

8.1 Introduction

The integration of photovoltaic (PV) systems into the electric grid has gained significant attention due to the growing demand for renewable energy sources. Solar energy, being abundant and environmentally friendly, is a key component in the transition toward a more sustainable energy infrastructure. PV systems convert sunlight directly into electricity using semiconductor materials, providing clean energy with minimal carbon emissions.

Grid integration of PV systems presents both opportunities and challenges. On the one hand, PV systems can significantly reduce the dependency on fossil fuels, contribute to energy security, and reduce greenhouse gas emissions. On the other hand, the intermittent and variable nature of solar energy can pose challenges to the stability and reliability of the electric grid. Addressing these issues requires advanced control systems, energy storage solutions, and regulatory frameworks that allow for seamless integration.

As PV installations increase globally, utilities and grid operators must adapt to the decentralized nature of energy generation. Traditional grids, designed for centralized power generation, now face the challenge of accommodating distributed generation (DG) sources like PV systems. This requires enhanced grid infrastructure, real-time monitoring, and grid balancing techniques [1].

8.2 Grid-Connected PV Power Systems

Systems for producing solar electricity can be broadly divided into two categories: grid-connected and standalone. The grid-connected system is covered in this section. Figure 8.1 shows the block diagram of a PV system connected to the grid.

Concentrated Solar Power Systems, First Edition. Bellamkonda Pragathi and D. P. Kothari.

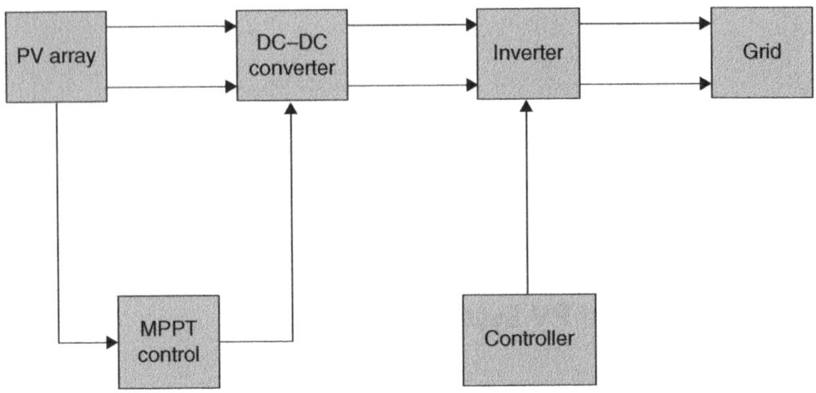

Figure 8.1 Block diagram of grid-connected PV power system.

The PV array, DC–DC converter, inverter, and related controllers make up the grid-connected system.

In the near future, grid-connected PV systems will become much more prevalent. Utility companies may be worried about the effects of fluctuating solar generation on the quality of the power, the low-tension (LT) distribution grid, and worker safety, though, as the grid was not specifically built for large-scale DG. As a result, integrating distributed energy sources like solar PVs presents several difficulties and problems. This chapter discusses the problems and obstacles as well as potential fixes.

Thanks to enabling laws and regulations, distributed kW-size solar PV plants—mostly rooftop models—are only now starting to be installed. Because of this, the majority of utility providers have little expertise in managing dispersed PV systems that are primarily connected to the LT distribution grid. Since they are in charge of ensuring a steady supply of electricity for the grid, they have some legitimate worries about distributed solar PV:

- **Power quality**: Concerns over the quality of the power being introduced into distribution networks are held by distribution corporations (DISCOMs). The main causes of this include DC injection, harmonics, and flicker [2].
- **Safety**: Utility companies have legitimate concerns regarding the security of their personnel, particularly when navigating the potential for inadvertent island formation as a result of the operation of distributed solar PV systems.
- **Low-voltage distribution system**: They are also worried about how the widespread use of dispersed solar generators will affect the LV distribution grid (voltage levels, power factor, increased equipment wear and tear, etc.).

- **Costs associated with transactions**: The noticeably increased transaction effort associated with metering, inspections, and certificates is another logistical concern for utility businesses.
- **Quality of power and inverter**: The core component of a solar PV system that guarantees grid integration and power quality is the inverter. The quality of the power fed into the grid can be influenced by three key technical parameters, which are DC injection, flicker, and harmonics. Associated Institute of Electrical and Electronics Engineers (IEEE) standards may be able to address the problems.
- **The CEA, or Central Electricity Authority**: IEEE 519 is the standard for harmonics, where the total harmonic distribution (THD) level must be less than 5%. IEEE 61000 is the standard for flicker. IEEE 1547 is the standard for DC injection, where the highest allowable level for the full-rated output at the interconnection point is 0.5%.
- **Range of operation of inverter**: The solar system is dependent on the quality of the grid and can only operate within a specific voltage and frequency range, in addition to the three fundamental criteria already mentioned. These ranges differ between nations. The system may currently operate within the 47.5 and 51.5 Hz frequency bands and between 80% and 110% of the nominal voltage, according to the CEA (for India).
- **Support from inverters to the grid**: Modern inverters can help with fault recovery and enhance certain grid characteristics in addition to integrating with the distribution system and preserving the power quality required by it. In the event of brief grid faults or breakdowns, inverters can maintain grid connectivity using the low/high-voltage ride-through (LHVRT) and low/high-frequency ride-through (LHFRT) functionalities. Reactive power support and power frequency droop are two grid-supporting tasks that inverters can provide to help keep local grid parameters within their normative bounds, sometimes more affordably than with centralized solutions. Inverters with this kind of intelligent functionality typically don't require any additional costs, and integrating and activating it doesn't pose any technical difficulties. The majority of these features are already available to a sizable portion of the Indian string inverter market.

Many of the previously stated features are likely to be incorporated in the upcoming revision of IEEE 1547, the reference guideline for DG. Some of these features are already available in a draft form (IEEE 1547a). In certain nations, more sophisticated features are being explored, some of which could need communication between the inverter and utility/energy markets. These features include the ability to adjust default settings in response to shifting grid conditions, to support connect/disconnect instructions, to limit the maximum active power upon utility company instruction, and more.

- **Inverter and safety**: Islanding is the state in which a distributed solar system keeps the circuit powered even in the absence of utility company grid electricity. Utility personnel may not be aware that a circuit is still powered while performing maintenance or repairs, which puts them at risk of islanding. The anti-islanding functionality, which requires the PV system to stop energizing the grid as soon as grid power is absent, is mandated by most nations, including India, to protect safety. "Intentional islanding," a newly developed feature, enables distributed solar PV to sustain some loads during grid outages while taking appropriate safety precautions. In India, where load shedding and power outages are frequent in the majority of DISCOM service areas, this is extremely pertinent [3].

8.3 Inverter Control Algorithms

The classic configuration of a grid-connected solar PV system consists basically of two components. A PV array serves as the energy supply for the first stage. The energy accumulated in an intermediate-storage element (DC-link capacitor) is converted and adjusted by an inverter at the output stage to meet the utility grid's specifications for voltage and power quality. The traditional two-level inverter is commonly utilized as the interface between the grid and DC link in low-power solar systems. In applications requiring high power and medium voltage, multilevel converters are frequently employed.

Over the past few decades, a number of control strategies have been put out for grid-connected applications. Stability, little harmonic content, quick dynamic response, and simplicity are the goals of the controllers. Trade-offs are typically necessary, though. For instance, the intrinsic stability and quick and reliable dynamic response of the classical hysteresis current controller are compromised by phase interference and fluctuating switching frequencies in isolated neutral three-phase systems. The voltage-oriented control method overcomes these shortcomings by controlling the load currents, which are divided into direct quadrature (d-q) components in accordance with the synchronous reference frame (SRF) theory. This results in a constant switching frequency and, as a result, a well-defined harmonic spectrum. Its dynamic performance and stability, however, are heavily reliant on load conditions. A new class of control systems has been made possible by the rapid development of powerful microprocessors, which offer advantages including flexibility and the capacity to conduct intricate online computations. These advantages are used by the predictive control technique to determine which control action, out of all the options, would best meet predetermined criteria. However, there is a chance that different load parameters will affect performance. Direct power control is another new control technique that chooses the appropriate switching state by using a switching table and online estimates of control variables. This method's direct regulation of the instantaneous active and reactive powers has demonstrated its great suitability for grid-connected applications [4].

8.4 Synchronous Reference Frame-Based Current Controller

Only the active portion of grid current needs to be injected by the DC–AC stage. Therefore, for every grid frequency, the steady-state current error between the intended and actual grid current should stay zero. The phase-locked loop (PLL) creates unit voltage templates (sine and cosine components) by tracking the phase of the input voltage signal. High-frequency harmonic components are removed by filtering the d-q components of currents. Subsequently, the d-q frame is converted back into one-phase components. The final switching signals, or the pulses for the inverter, are generated by the hysteresis-based pulse width modulation (PWM) signal generator by comparing this current with the source current and feeding it any mistake between them. Figures 8.2 and 8.3, respectively, show the reference current extraction using the SRF theory and the inverter control utilizing the PI controller [5].

These reference currents are compared with the actual current, and errors between both will be the required amount of reactive current supplied by the inverter. The error of the current is used to generate pulses for the inverter.

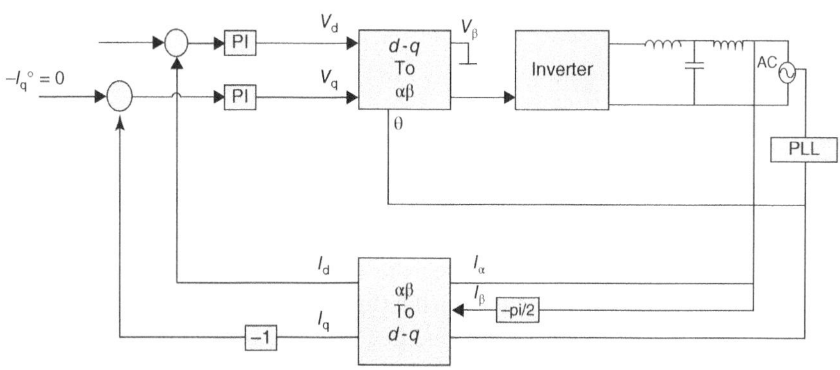

Figure 8.2 Reference current extraction using synchronous reference frame (SRF) theory.

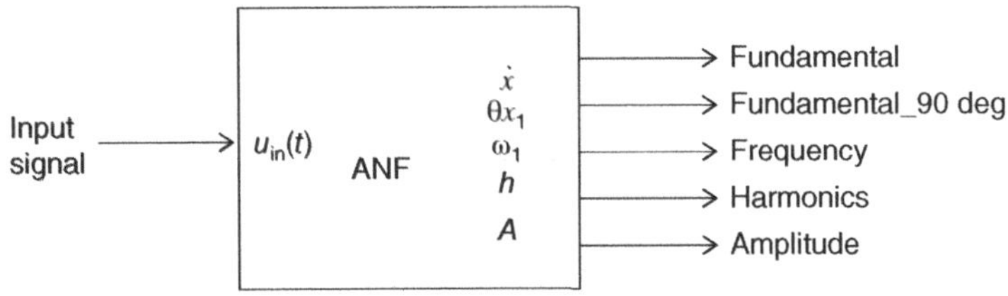

Figure 8.3 Basic block of adaptive notch filter.

8.5 Digital PI-Based Current Controller

Digital PI control can be utilized to regulate the different inverter parameters, like voltage and current. The controlling algorithms can be programmed into microprocessors to integrate the digital PI control into the system hardware.

The controller uses the PI algorithm to reduce the error between the inverter current and reference current or to update the response. The process of detecting and translating grid voltage into current results in the reference current, which is a function of grid voltage. By doing this, the inversion output current is kept in phase with the grid voltage and at the unity power factor. The instantaneous error between the reference current and the inverter output current is monitored by the PI controller. Ultimately, the PWM signal of a single-phase inverter is obtained by comparing this error signal with a carrier signal. The following formula represents the PI controller algorithm: While integral gain necessitates approximations, a discrete time domain equation for percentage gain can be applied directly, just like it was in the continuous form. Hence, the time domain form can be obtained by applying a technique known as the trapezoidal sum approximation.

8.6 Adaptive Notch Filter-Based Grid Synchronization Approach

Both linear and nonlinear forms of loads can have their harmonic current and reactive power compensated for using the adaptive notch filter (ANF)-based method. A PI controller is used to keep the DC-link voltage constant.

To preserve the amplitude, phase, and frequency parameters in power quality enhancement, the interfaced PV system is synchronized with the grid using the ANF technique. The periodic form of the signal is described as follows:

8.7 Modeling, Simulation, and Hardware Implementation of Controllers

This section discusses the modeling and simulation of grid-connected PV systems that have been done in the MATLAB® environment. For grid-connected PV systems, a control method based on ANFs is also suggested for the voltage source converter (VSC). With both linear and nonlinear loads, the suggested control method offers improved reactive power and harmonic compensation. A PI controller is used to maintain a steady DC-link voltage. Using MATLAB's Simulink toolbox and the SimPower System, an entire system has been modeled and simulated. To guarantee a consistent power supply even in the face of significant disruptions and fluctuations, a hardware configuration and controller for the integration of PV systems into distribution networks has been created. The constructed experimental setup is used to validate the suggested controller's outcomes. Figure 8.4 shows the completed MATLAB Simulink model for a three-phase grid-connected system. Figures 8.5 and 8.6 show the results of an analysis of the system's performance using various parameters, including the DC power supplied by the PV system, the DC-link voltage, the three-phase voltage, and the active and reactive power supplied by the system at different meteorological parameters.

PV voltage and current are crucial factors that determine how well a PV system works.

The MPPT control algorithm uses these parameters as an input to generate pulses for the DC–DC boost converter. Maintaining a desired DC link voltage requires the usage of a DC–DC converter. The DC-link voltage is kept constant by using a large-value capacitor. The boost converter maintains the DC connection voltage steady at about 700 V. For IGBT switches, the inverter control block

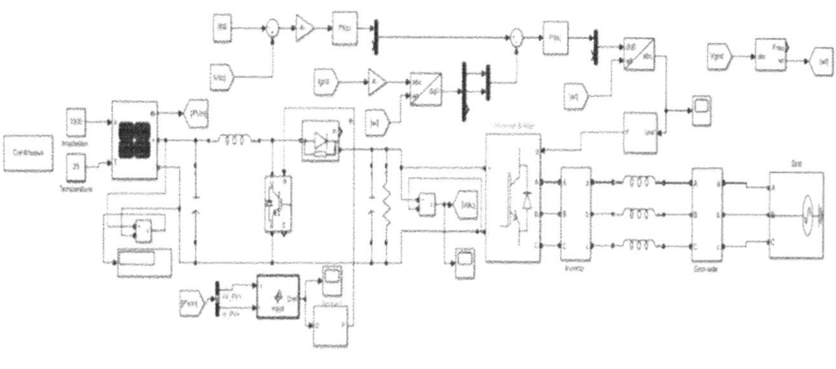

Figure 8.4 MATLAB® simulation of three-phase grid-connected PV system.

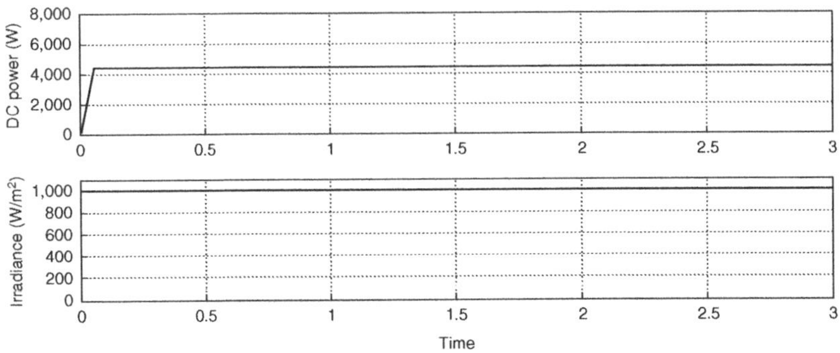

Figure 8.5 DC power of PV system at constant irradiance.

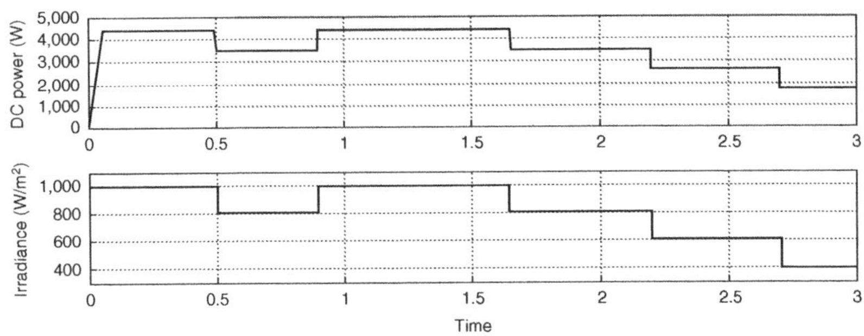

Figure 8.6 DC power of PV system at varying irradiance.

generates pulses. Figures 8.5 and 8.6 clearly illustrates the current harmonics produced by the solid-state switches used by the VSI. Illustrate how the irradiance parameter affects the active power while the meteorological parameter variation has no effect on the nearly constant reactive power supply. The PV system only provides the partial amount of reactive power needed. The load current's THD and its effects at various penetration levels are examined and shown in Figures, respectively. Additionally, Figure presents the analysis of the effects of growing PV penetration on the current grid using THD and voltage at the PCC point [6, 7].

We may infer from the data that the suggested controller performs satisfactorily and that every result falls within the intended ranges.

A grid synchronization technique based on an ANF for a PV system, along with MPPT techniques, has been developed. In this approach, a reference phase signal synchronized with the grid voltage is provided to standardize the system with grid codes and power quality standards. Therefore, this approach plays an important role in the grid-connected PV system. Because the output of the PV array fluctuates based on meteorological parameters such as solar irradiance, temperature, and wind speed, in order to maintain a constant DC voltage at VSC input, MPPT control is required to track the maximum power point from the PV array. In this work, a variable step size P&O (perturb and observe) MPPT technique with DC–DC boost converter has been used at the first stage of the system. The proposed algorithm divides the dP_{pv}/dV_{pv} curve of the PV panel into three separate zones (i.e. zone 0, zone 1, and zone 2). A small tracking step size is used in zone 0, whereas zone 1 and zone 2 require a large step size in order to obtain a high tracking speed [8, 9]. Further, the ANF-based control technique is proposed for the VSC in a PV generation system. This approach is used to synchronize the interfaced PV system with the grid to maintain the amplitude, phase, and frequency parameters, as well as power quality improvement. This technique offers the compensation of harmonics current and reactive power with both linear and nonlinear loads. To maintain a constant DC-link voltage, a PI controller is also implemented and presented in this work. The complete system has been designed, developed, and simulated using the SimPower System and Simulink toolbox of MATLAB.

The suggested system's performance under both linear and nonlinear loads is tested and shown in Figures 8.7 and 8.8. For simulation investigations, a 50 kW linear load with a trailing power factor is employed. The PV array is planned to have a maximum power output of 20 kW. To provide the load with the maximum

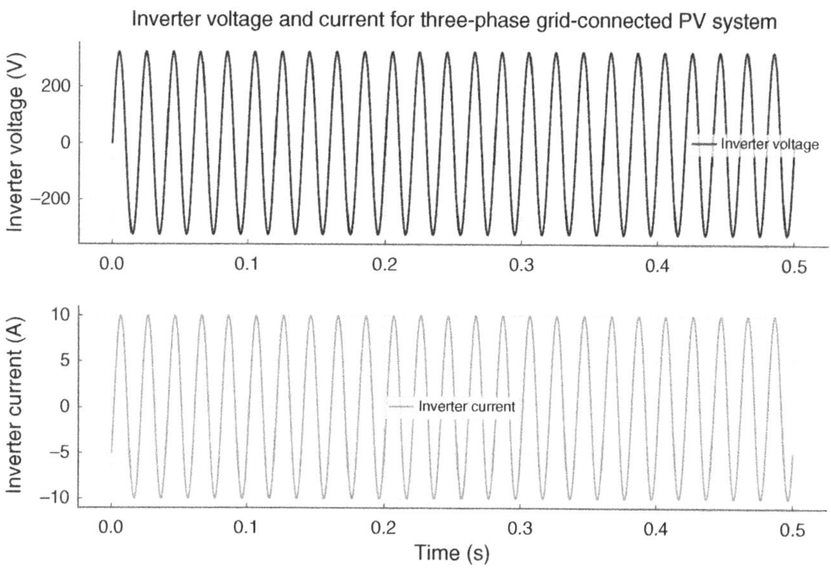

Figure 8.7 Inverter (a) voltage and (b) current.

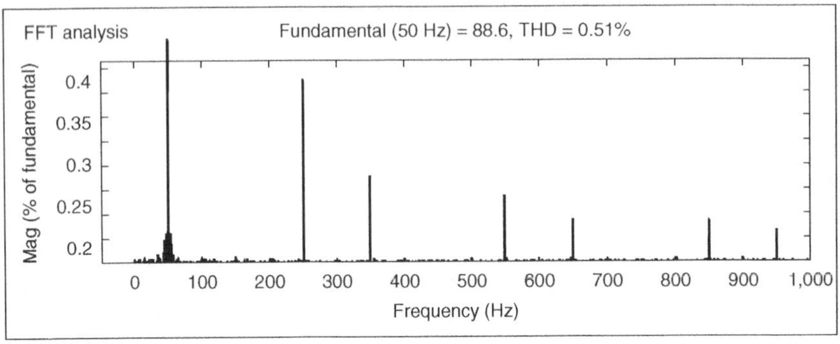

Figure 8.8 THD of load current.

amount of power, the system employs MPPT control. The DC-link voltage reference value is maintained at 700 V. The DC-link voltage is adjusted by a PI controller to its reference value. The waveform makes it evident that at first, the source alone is feeding the entire load current; however, after the inverter is switched on, the load current is shared by the PV system and the grid. Figure 8.9 shows the system's performance under a linear load. An adjustment to the PV array's irradiance is also implemented in order to verify the precision and effectiveness of the suggested MPPT method. The irradiance increases to 1200 W/m^2 at 0.1 second and decreases to 800 W/m^2 at 0.2 second. The PV array feeds a portion of the load's active power for up to 0.35 second. The inverter provides complete reactive power correction, meaning the source does not deliver reactive power to the load. The proposed system operates at unity power factor because the grid currents and PCC voltages are in phase [10] (Figures 8.10 and 8.11).

As a result, after 0.35 second, the solar irradiation is lowered to zero and the PV current also drops to zero. The VSC provides the necessary adjustment for this mode of operation. It is evident from the simulated results that the system performs satisfactorily under linear load circumstances. The different parameters of the suggested system under a nonlinear load condition are displayed in Figures 8.11. It is evident that the grid operates at unity power factor when the rents are in phase with the PCC voltages. With the VSC's compensation, the grid currents are balanced and sinusoidal regardless of the nonlinear load. When the sun intensity drops to zero at 0.5 second, the VSC takes over and compensates the loads. In this instance, the PV array has once more had its irradiance altered. It is evident from the output waveforms that, when using the suggested control technique, the system operates effectively even for nonlinear loads.

Figure 8.9 THD of grid at (a) 10%, (b) 15%, (c) 20%, (d) 25%, and (e) 30% PV penetration.

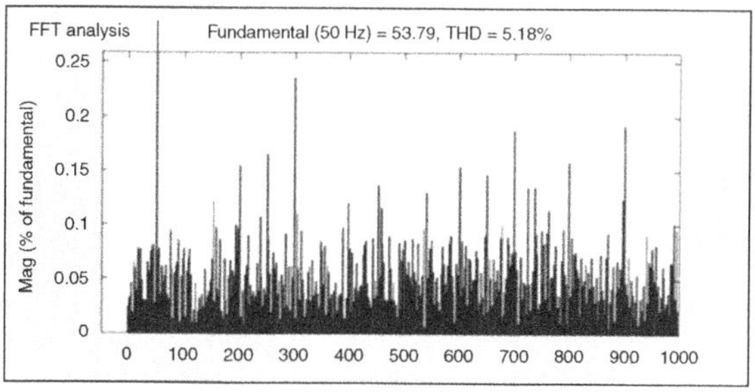

(a)

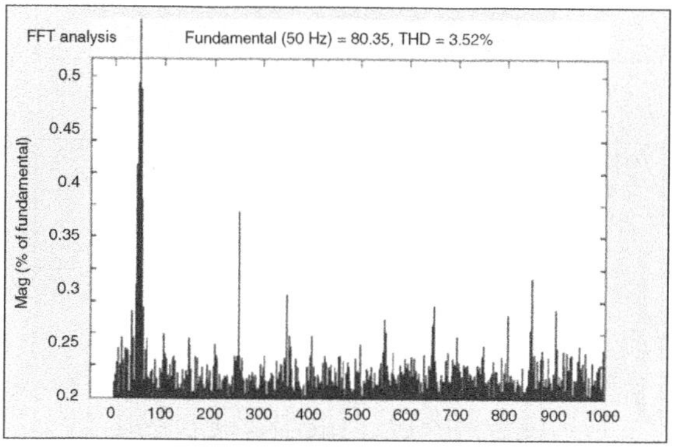

(b)

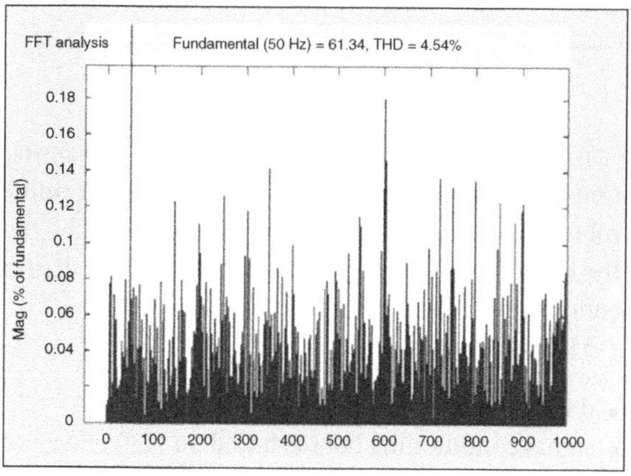

(c)

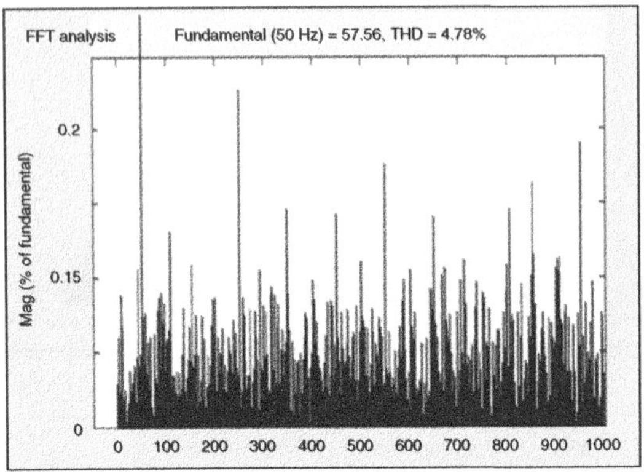

(d)

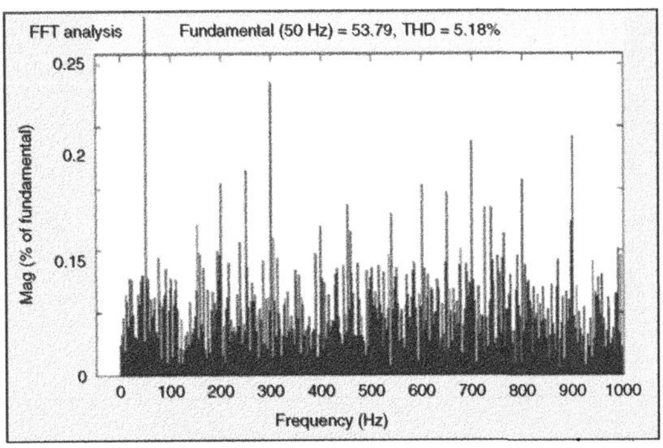

(e)

Figure 8.9 (*Continued*)

To guarantee a steady power supply amid significant disruptions and fluctuations, hardware engineers can create a power conditioning device. The controller's robustness against intermittency is confirmed by the results, which also improve the controller's degree of integration into the distribution system. The following components of the experimental setup are

5 kWp PV array;

- dSPACE DS1104 R&D controller board;
- signal conditioning boards based on Hall effect voltage and current sensors;
- power conditioning unit consisting of
  - o IGBT-based converter;
  - o inverter/MPPT, gate drivers, and
  - o optocoupler integrated circuits (ICs);
- isolation transformer; and
- power quality analyzer.

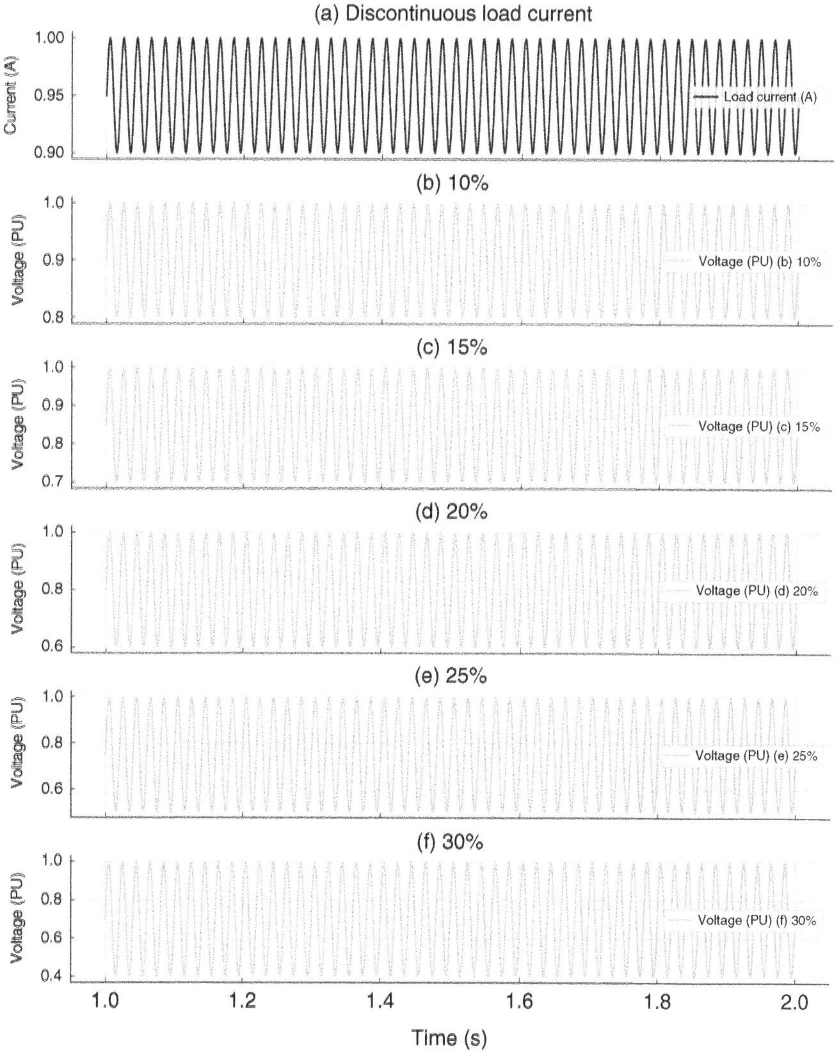

Figure 8.10 (a) Discontinuous load current and voltage fluctuation levels at (b) 10%, (c) 15%, (d) 20%, (e) 25%, (f) 30%.

8.8 Conclusion

The integration of PV systems into the electric grid represents a pivotal step toward achieving sustainable and cleaner energy systems. As the world shifts toward renewable energy to combat climate change, PV systems offer a reliable,

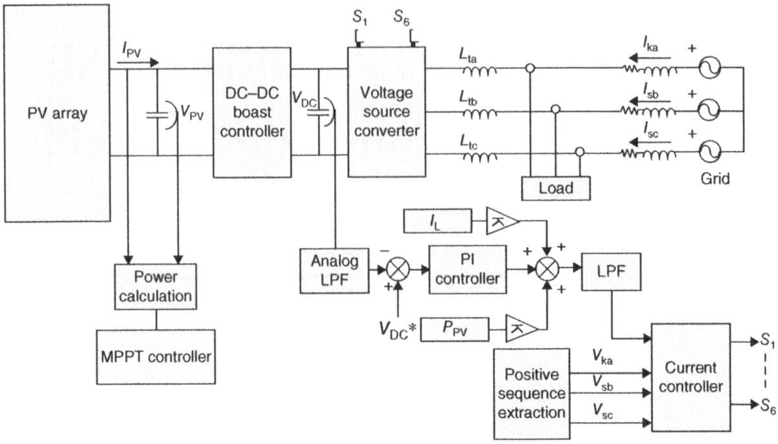

Figure 8.11 Block diagram of system under study.

scalable, and environment-friendly energy source. However, the successful integration of PV into the grid comes with challenges such as intermittency, grid stability, and energy management.

Technological advancements, such as smart grids, energy storage systems, and real-time monitoring, have proven essential in overcoming these challenges. By improving grid flexibility and incorporating robust energy management strategies, PV systems can be efficiently integrated, enhancing grid reliability while reducing dependence on fossil fuels.

In conclusion, with the right policies, technological innovations, and cooperative frameworks, the grid integration of PV systems will play a crucial role in shaping a sustainable and resilient energy future. This transformation not only addresses environmental concerns but also opens up new opportunities for economic growth and energy independence.

References

1 Suleiman, M.A. and Ahmed, A. (2024). Analysis of grid-tied solar photovoltaic energy generation under uncertain atmospheric conditions using adaptive neuro-fuzzy control system. *Archives of Advanced Engineering Science* https://doi.org/10.47852/bonviewAAES42022110.

2 Kiasari, M., Ghaffari, M., and Aly, H.H. (2024). A comprehensive review of the current status of smart grid technologies for renewable energies integration

and future trends: the role of machine learning and energy storage systems. *Energies* 17 (16): 4128.

3 Maka, A.O. and Chaudhary, T.N. (2024). Performance investigation of solar photovoltaic systems integrated with battery energy storage. *Journal of Energy Storage* 84: 110784.

4 Khalid, M. (2024). Smart grids and renewable energy systems: perspectives and grid integration challenges. *Energy Strategy Reviews* 51: 101299.

5 Boscaino, V., Ditta, V., Marsala, G. et al. (2024). Grid-connected photovoltaic inverters: grid codes, topologies and control techniques. *Renewable and Sustainable Energy Reviews* 189: 113903.

6 Forrousso, S., Kaitouni, S.I., Mana, A. et al. (2024). Optimal sizing of off-grid microgrid building-integrated-photovoltaic system with battery for a net zero energy residential building in different climates of Morocco. *Results in Engineering* 22: 102288.

7 Saleem, M.I. and Saha, S. (2024). Assessment of frequency stability and required inertial support for power grids with high penetration of renewable energy sources. *Electric Power Systems Research* 229: 110184.

8 Modi, G. and Singh, B. (2024). Power quality improvement in solar energy conversion system integrated to weak AC grid. *Journal of The Institution of Engineers (India): Series B* 1–16.

9 Mukherjee, S., Mandal, R., and Chatterjee, S. (2024). Comparison of harmonics mitigation techniques for grid-connected PV System and Introduction of a Concept of Hybrid Filter. *Applied Solar Energy* 60 (1): 138–148.

10 Wen, X., Shen, Q., Zheng, W., and Zhang, H. (2024). AI-driven solar energy generation and smart grid integration a holistic approach to enhancing renewable energy efficiency. *International Journal of Innovative Research in Engineering and Management* 11 (4): 55–66.

9

Optimization of Concentrating Solar Power (CSP) Plant Designs Through Integrated Techno-Economic Modeling

9.1 Introduction

Since simulation aids in the prediction of the economic, energy, and operational features of an actual plant installation, it is crucial for supporting decisions pertaining to investment in and design of concentrating solar power (CSP) plants. This chapter examines accepted methods for CSP plant modeling and design optimization. A novel approach to solar thermal power plant design optimization combines a multiparameter optimization algorithm that enables the optimization of the pertinent plant design parameters with the energetic and economic simulation of a whole power plant at the component level. This concept is applied to a state-of-the-art parabolic trough plant with a two-tank molten salt storage to demonstrate the process and the possible outcomes. Still, the methodology and even the majority of the findings hold true for various CSP systems.

9.2 The Most Recent Advancements in CSP Plant Design and Simulation

Various software tools and approaches are available now for simulating CSP facilities, depending on the simulation's scope. The ensuing sections provide an outline of the various methods.

9.2.1 Calculating Energy Yield

The unstable solar resource is the primary distinction between solar thermal power plants and conventional power stations when calculating energy yield. The sun's location in relation to the collector orientation affects the collector efficiency, which in turn affects the thermal output of the collector field, even in the absence of clouds that would otherwise lower the amount of available direct

Concentrated Solar Power Systems, First Edition. Bellamkonda Pragathi and D. P. Kothari.

solar irradiation. Therefore, as is frequently done for conventional power plants, calculating a mean annual energy yield based on an expected number of yearly full load hours is insufficient.

There are techniques for carrying out short-term dynamic simulations with high time resolution (e.g., time-step one second) for the energetic simulation of solar thermal power plants. These instruments concentrate on comprehending plant dynamics and creating collector and/or power plant control parameters. ColSim from Fraunhofer ISE and the German Aerospace Centre's (DLR) application library DissDyn are two examples of dynamic simulation tools. A comparative description of both programs for usage in direct steam-generating line-focusing collectors. To accurately describe transient flow effects, the tube's flow is divided into small control units or tube segments. Thus, it is possible to mimic dynamic effects such as momentarily different mass flows at the collector inlet and outlet as a result of density variations, which aids in the development of effective management strategies. High spatial and temporal resolution is necessary to adequately describe these effects, which leads to a computation time that is frequently longer than real time (i.e. it takes longer to simulate an hour of plant operation than it does in real time).

Consequently, "pseudo steady-state" calculations—which are based on hourly discretization with simplified consideration of dynamic effects—are used to calculate the energy yield of representative periods of one year or several years. Currently, there are essentially two ways to go about doing this:

1) The method utilized by CSP research institutes is based on a fairly thorough simulation of the solar block, including the solar field and storage (if applicable), as well as characteristic curves that characterize the power block for the thermal to electric power conversion ratios. Some examples are the Greenius from the DLR and HTW Berlin, the System Advisor Model (SAM) from the National Renewable Energy Laboratory (NREL), and a version of ColSim from Fraunhofer ISE that simplifies the dynamics. ColSim was initially created to mimic solar thermal systems' dynamic behavior and create control schemes for them. ColSim's functionality was expanded to include the simulation of solar thermal power plants energetically starting in 2002. However, this extension was based on a simplified analysis of dynamic effects, which did not take plug flow into account.
2) Software vendors that specialize in modeling conventional thermal power stations have begun to extend their modeling skills to include CSP facilities in the past two years. As of right now, the state of the art indicates that all of these tools have two things in common: they compute the complete thermal cycle of the CSP plant for each time step, and because of their historical background, they offer detailed power block models. These tools have a sluggish computing speed since they require the solution of complex equation systems using

thermodynamic calculations. Modern PCs require hours to do a single annual yield calculation (hourly resolution), yet the aforementioned tools, SAM, Greenius, and ColSim, only require a few seconds.

Under the auspices of IEA-SolarPACES, an initiative to standardize energy yield calculations for CSP facilities internationally has begun. The United States' NREL hosted a workshop in March 2010 on modeling the annual performance of solar thermal power plants. Various methods were discussed, and a roadmap for comparison calculations and program validations.

All of the tools compute the energy yield along the energy conversion chain, including thermal power, optically absorbed solar power, direct solar irradiation, electric gross power, and electric net power, despite variations in the models employed by each tool [1–3].

9.3 Economic Simulation

Certain modeling techniques solely concentrate on the plant's energetic simulation. Some incorporate an economic model as well. The economic models are different based on how big the simulation is. The levelized cost of energy, or LCOE, can be used to assess the effects of technical differences in most circumstances. Other methods incorporate a thorough cash flow analysis of a plant, taking into account the cost of capital (debt and equity financing), time-dependent electricity sales prices, and energy-political incentives such as feed-in tariffs or tax credits. Every commercial player in the solar thermal power industry, including project developers, general contractors for plant construction, plant operators, and original equipment manufacturers, has an economic model that they use to assess a project's financial viability and profitability in accordance with their particular business procedures.

9.4 Solar Thermal Power Plant Design Procedure

These days, several successive computations are carried out to determine the design of CSP plants. Typically, several tools are used to estimate various technical subsystems or plant economics. Parameter studies are used to vary and optimize the parameters of interest using the tools that represent a particular aspect or subsystem of the CSP plant. At the system level, solar field size and storage size are the primary parameters that are often optimized with regard to techno-economic criteria.

One tool that combines the energy and economic simulation of a CSP plant is the previously described SAM. Because of the US Department of Energy's

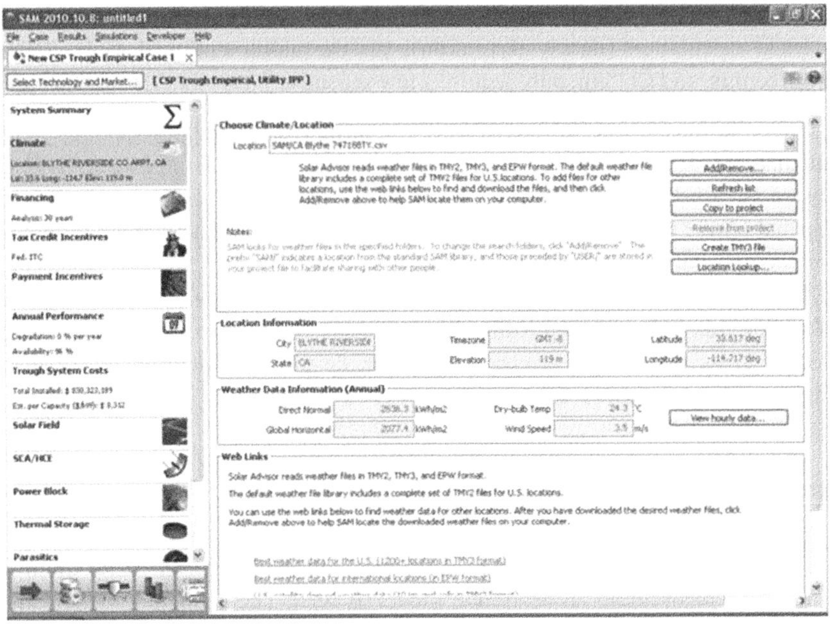

Figure 9.1 Screenshot of NREL's System Advisor Model for simulating CSP plants energetically and economically.

generous financing over the past few years, SAM has undergone significant evolution. Probably, the most popular CSP simulation tool (for plant yield and plant economics) is SAM. This is true not only for a growing number of well-known CSP actors (like technology providers and engineering consultants) but also for less technically inclined CSP actors like project developers or financial industry players. SAM provides a user-friendly graphical user interface (GUI) and a wide range of predefined technical options for the user to choose from (and optionally adjust) (see Figure 9.1).

SAM uses hourly weather data to determine the plant's annual and hourly energy yield based on site and technological criteria. SAM can be used to assess projects economically using cash flow analysis or levelized cost of energy (LCOE) in addition to energy data. The cost of the investment, the cost of operation and maintenance (O&M), the lifespan of the plant, debt and equity financing, the cost of fuel and electricity, fiscal factors, and other factors are some of these economic performance requirements. By changing the parameters of the plant design, a fundamental optimization feature facilitates the search for parameter configurations that maximize an objective function, such as LCOE [4].

One disadvantage of SAM is that it does not use a physical power block model; instead, power block efficiency calculations are dependent on static parametric power block equations. The SAM default power block data needs to be critically revised and/or generated on a case-by-case basis using other simulation tools that are dedicated to power block simulation (e.g. Thermoflex or Ebsilon; see above). This is because power block integration is a very project-specific issue (depending on plant size, availability of cooling water, power block integration concept of solar field, and optional co-firing).

The subsequent sections outline a strategy to address the aforementioned shortcomings of the current tools, including the inability to efficiently solve the multidimensional optimization problem when designing CSP plants and the laborious process of generating power block data in conjunction with slow power block simulation.

9.5 Multivariable Optimization of Concentrating Solar Power (CSP) Plants

Today, plant designs are simulated and optimized utilizing various tools and techniques, because numerous design parameters can be altered and adjusted for each project, designing CSP facilities is a multivariable optimization process. However, there hasn't yet been a technique or piece of software that combines the interdependent impacts of solar blocks and electricity blocks at the system level. Therefore, evaluating the mutual interdependencies is difficult.

A new approach has been proposed to integrate techno-economical modelling and optimization (including power block optimization) and the key components of this technique are outlined in the following sections. This method can assist in finding a well-designed plant concept that is specific to a project rapidly. To promote sales, for example, this enables project developers to swiftly arrive at a sound project design and project evaluation. This avoids person-months of concept engineering or, in the event that it is not possible, the use of less-than-ideal (and costly) plant designs. Furthermore, based on a top-down strategy in which the entire plant design is optimized first, such a tool can offer requirements for precise component engineering at an early planning stage.

The goal of designing a commercial power plant is always to identify the most economical plant design possible given a specific set of constraints. Economic performance metrics, such as the net present value and the levelized cost of energy (LCOE), will be considered by decision-makers.

In the sense that a power plant must be developed in accordance with the particular requirements, such as solar site conditions, cooling water availability, nominal power, and rules for fossil co-firing, technical design determination for CSP facilities is project-specific. The subsystems, such as the solar field, power cycle, condenser, and, if applicable, heat storage and/or boiler for co-firing, must be planned and dimensioned based on these parameters. Each of these elements can be viewed as a variable in an optimization problem with multiple parameters. These design characteristics affect energy yield as well as cost. Relevant interdependencies are explained using a few instances [5].

- A larger solar field increases the number of full load hours for a particular power block size, but it also raises investment and O&M costs.
- While higher operating temperatures improve the thermal cycle's efficiency, they also increase heat losses in the solar field.
- The condenser's large heat transfer surfaces will increase cost and efficiency. Condenser design will be impacted by the solar field's cost, however. Because the solar heat is particularly expensive due to the high investment requirements for the solar collector, condensers in CSP facilities are probably larger than condensers for conventional power plants of the same size.

These days, a technical design concept is frequently produced before an economic evaluation of a plant concept is completed. When these stages are followed in order, the following issues occur:

- When it comes to optimizing a plant layout in light of economic decision criteria, various instruments—and perhaps even distinct individuals, including financial specialists and specialized engineers—are needed. Design computations and evaluations carried out in iterative loops are costly and time-consuming.
- Given the large number of variables that influence one another, it is possible that significant optimization potential goes undiscovered.
- Professionals with expertise in financial analysis and engineers with varying specializations, like solar field or power block, run the risk of focusing only on their specific targets and neglecting the plant's overall cost performance.

This author and colleagues have developed the new package OPTISIM. To undertake design optimizations, it intends to combine the benefits of the precise power block modeling tools discussed above with the speed of annual yield calculations. Thermoflex (in conjunction with the cost estimation feature PEACE) and ColSim (for the energetic simulation of the solar field and thermal storage) are the components of the OPTISIM package. Other software tools, most of which are freely available, are used for multiparameter optimization, database communication, and economic plant modeling.

The new methodology takes into account all pertinent site-specific, technical, and economic elements at the same time, hence assisting in the complete exploration of optimization possibilities throughout the plant design process. It is possible to calculate the electricity yield particular to a location and design using this comprehensive methodology. Financial performance metrics like LCOE and investment expenses are computed concurrently.

This program is used in conjunction with an appropriate, time-efficient optimization method to optimize multiple design and operating parameters that are impactful on the energetic and financial quality of the system concurrently. Specific requirements can be utilized to determine the amount of variable factors in a simultaneous optimization as well as the quality function, such as the LCOE. Such factors include, but are not limited to, temperature and pressure levels; solar field size; power block configuration; condenser size; condenser part load operation; and solar field design parameters.

Power plants with parabolic trough and linear Fresnel collectors, hybrid and solar-only plants, plants with or without thermal storage, and plants for direct steam generation or the use of a heat transfer medium can all benefit from the modeling and optimization model [6, 7].

Figure 9.2 shows the methodology that is used in this new approach. First, a set of technical and economic parameters has to be chosen to mimic the layout of a power facility. Site of the power plant, power block, collector, storage (if applicable), and other economic presumptions are all mentioned in the parameters. Certain characteristics, including location and irradiation statistics, are constant and unaffected by plant design. An evolutionary algorithm simultaneously optimizes the variable parameters in relation to the goal function.

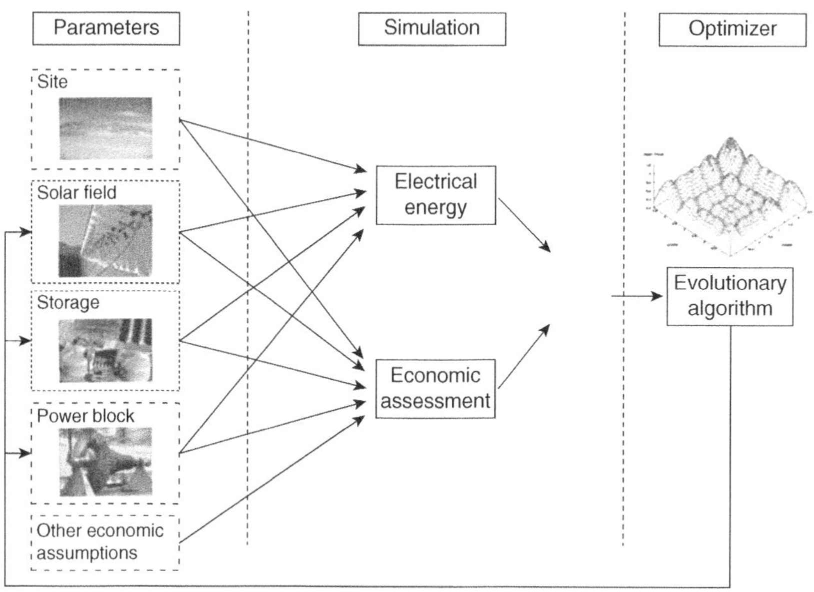

Figure 9.2 General structure of the techno-economic system simulation and optimization model. Source: Illustration arrangement: author, single photos/graphs from top to bottom, left to right: Atacama Desert [8], Relloso and Delgado [9], Strelow [10]. Source for Storage and Power block images: Morin [11]/with permission of Elsevier.

The variables are determined by the optimization needs; the size of the solar field is one example of a variable parameter. Based on hourly data, the yearly energy yield for a sample time period (such as a year) is computed using all of these assumptions. Unlike conventional power plants, the sun's constantly shifting location and the amount of solar radiation necessitate precise annual energy yield estimations. The mean yearly cost is computed based on the economic assumptions. The integration of technical and economic simulation outcomes yields an objective function, such as the LCOE, that signifies the system layout's economic quality.

After that, a multiparameter optimization technique is applied to maximize the goal function while optimizing the pertinent plant parameters. The optimizer addresses both the plant's structural problems (such as various process designs with varying numbers and arrangements of feedwater heaters and reheat sections (0, 1, or 2 reheat sections) as well as steady-state operational characteristics (such as live steam temperature or live steam pressure). A GUI to adjust simulation parameters and track optimization progress has been developed (Figure 9.3).

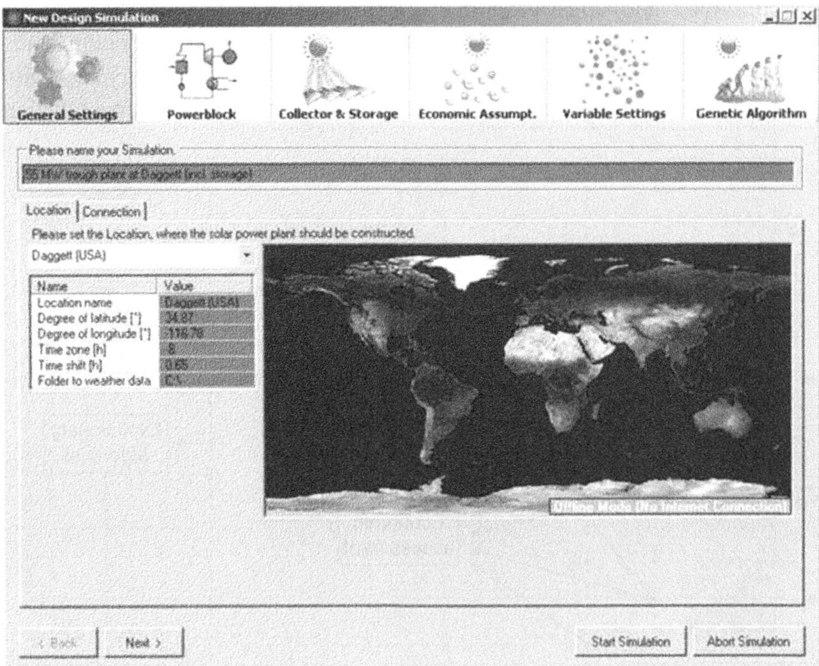

Figure 9.3 Graphical user interface for integrated simulation and optimization software package *OPTISIM*, aiming at integral techno-economic design optimization through fast but detailed power block modeling in combination with solar simulation and a multiparameter optimizer.

9.6 Overview of Optimization Methods

There are several different multiparameter methods available. To assess the applicability to CSP plant optimization, we have to look more closely at our optimization problem. The objective function, which in the example described below is the LCOE, has the following properties:

1) The LCOE function is nonlinear.
   Example: the solar field size influences the LCOE nonlinearly.
2) The LCOE function is not continuously differentiable.
   Example: the solar field size is defined by an integer number of collector loops with a standard length. This results in a step function.
3) The LCOE function may have several local minima.
   Reason: It cannot be ruled out that the LCOE has several local optima due to the interactions of technical and economical models. When optimizing several process structures at once the LCOE function has several local minima: e.g. a process with reheat (or even double reheat) will have higher optimal live steam pressures than processes without reheat.

Most methods for multiparameter optimization search for the nearest optimum, which is a local optimum but not necessarily a global optimum. Examples for such methods are the Simplex method, gradient methods, the Newton method, or the widely used Levenberg–Marquart Algorithm. Since such methods do not fulfill the above-mentioned third requirement (global optimization), they were not further considered in the context of this work. To date, there are no methods available that can ensure globally optimizing an objective function with the properties mentioned above. But there are several methods which in principle address the aim of global optimization, e.g. the multiplicator method of Lagrange (mathematical optimization), and the stochastic methods: Monte Carlo methods and evolutionary algorithms. Evolutionary algorithms conduct a stochastic but systematic search for improvement analogous to biological evolution. A configuration is characterized by a so-called individual. Several individuals form a population that evolves from generation to generation. Evolutionary algorithms have a big advantage over global optimization algorithms that the optimization and the model are separate which results in a clear interface which is easy to handle (more easy to handle than mathematical methods which often aim at differentiation of the objective function).

In the context of the *OPTISIM* software, an evolutionary algorithm (*GAlib* from *MIT*) was used for multiparameter optimization [12, 13].

9.7 Case Study Definition: Optimization of a Parabolic Trough Power Plant with Molten Salt Storage

In this section, the techno-economic modeling and optimization of a typical case of commercial interest is presented: a 50 MW_{el} parabolic trough power.

9.7.1 Definition of Optimization Task

The case study examines a 50 MW_{el} parabolic trough power plant, similar to the *Andasol-1* power plant (see Figure 5.4). The reference plant uses thermal oil as heat transfer fluid (HTF) and a molten salt thermal storage. As opposed to the *Andasol* plants, an air-cooled condenser and a Californian site in the Mojave Desert are assumed (Figure 9.4).

The Andasol-1 plant is located near Guadix in Southern Spain and uses a thermal storage utilizing 28,500 tons of liquid salt equivalent to 7.5 hours of full load operation and a solar field with a total surface of 510,000 m^2. The power-block model is based on the specifications of an operating 30 MW_{el} CSP plant in California, but upscaled to 50 MW_{el}. The cycle has a live steam temperature of 370°C, a design condenser pressure of 0.08 bar, one reheat section, and six

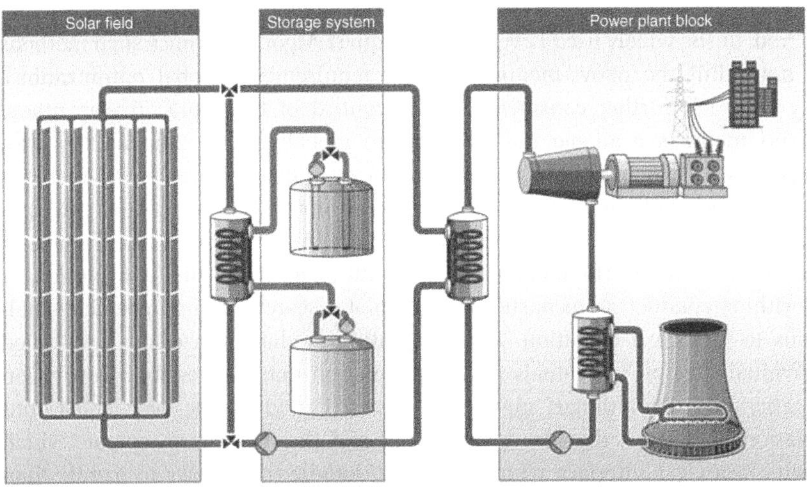

Figure 9.4 Sketch of the reference plant design based on the *Andasol-1* power plant in Andalucía, Spain. The thermal storage uses molten salt as storage medium, which is stored in a cold tank and/or a hot tank. Source: Sven Moormann [14]/with permission of Elsevier.

feedwater heaters. An air-cooled condenser instead of a wet cooling condenser has been used and no fossil co-firing was considered.

The reference design was then optimized for the site characteristics of Daggett in Southern California, USA (weather data retrieved) The main difference between the sites in Spain and California is the available solar irradiation, whereas the *Andasol* site has an average direct normal irradiance (DNI) of 2,136 kWh/(m^2a) and Daggett has 2,791 kWh/(m^2a), hence roughly 30% more solar irradiation [13].

Starting from the reference design, the following eight variables are simultaneously optimized, with the indicated purpose:

1) **Solar field size**: A larger solar field increases the power block full load hours but also increases dumping of thermal energy due to capacity limitations of the power block and of the storage (the so-called upper dumping).
2) **Distance between parallel collector rows**: Larger distance between parallel collector rows reduces mutual shading of collector rows and therefore improves optical efficiency. On the other hand, the cost of land of header piping increases with larger distance.
3) **Storage capacity**: Larger storage increases the operating hours of the power block and limits hours with low-efficiency part-load operation of the power block. But higher investment is needed. The size of storage is defined in tons of liquid salt.
4) **Upper solar field temperature**: Higher temperature leads to higher power block efficiency, but also solar field efficiency drops due to increased heat loss.
5) **Upper temperature difference in oil-to-water/steam heat exchangers**: The Californian reference plant assumes a terminal temperature difference (TTD) of 20 K between hot oil and hot steam. Smaller TTD can increase the efficiency of the power block because the live steam temperature rises (or reduces solar field temperature and hence heat loss). However, an argument in favor of a large TTD in the heat exchanger is that at a given live steam temperature, a large TTD will lead to low solar field oil inlet temperature and hence the HTF pumping power decreases.
6) **Live steam pressure**: For each process configuration, an efficiency-optimal live steam pressure relates to a given live steam temperature (examination of optimal pair of live steam temperature and pressure. Additionally, too high pressures must be avoided because of condensation at the outlet of the high-pressure turbine (danger of water droplet erosion).
7) **Reheat pressure:** With the herein assumed rule of equal temperature rise in all feedwater heaters, the parameter 'reheat pressure' determines not only the reheat pressure itself but also implicitly serves for calculating the final

feedwater temperature. Like the live steam pressure, the reheat pressure leads to a Carnotization of the heat cycle because both variables lead to a rise in temperature at which the (solar) heat is transferred to the water steam cycle. One side restriction is that the reheat pressure level must be chosen such that the steam dryness requirements are met at the outlet of the high-pressure steam turbine and of the low-pressure turbine (danger of water droplet erosion in final turbine stages). Final feedwater temperature and reheat pressure have an influence on lower oil temperature and hence on pumping power parasitics and energetic storage capacity.

8) **Size of air-cooled condenser**: Larger cooling systems lead to higher power block efficiency but also to higher investment cost. Large air-cooled condensers have high auxiliary power consumption. The optimal condenser size in the sense of a trade-off between cost and performance is searched for.

9.8 Applied Energetic and Economic Plant Models

The integrated techno-economic design optimization methodology can in principle be applied with any technical and economic models and assumptions. All models used in the calculations presented in this chapter allow for quantitative consideration of the effects.

9.8.1 Energetic Plant Model

The technical model of the plant describes the relevant energetic mechanisms in order to calculate the energy production of the plant under a set of side restrictions. The principle methodology is summarized as follows. The model used here is applicable for different time resolution steps but by default uses hourly resolution. For each hour of the year, the following steps of calculation are performed.

Based on the calculated sun position and the available DNI data, the optical collector performance is calculated. After subtraction of heat loss and thermal inertia effects—when heating up (or cooling down)—the thermal energy that the solar field can provide (potential) is calculated. First, the solar field performance is calculated surface specifically, referring to the collector aperture. The total field performance is calculated by multiplying the surface-specific performance by the total solar field surface. The total solar field surface in m^2 is rounded to an integer value of collector loops, which is the standardized collector module in a large solar field.

The HTF heated by the solar field is used to provide thermal energy to the power block or to load the storage, in case the power block already runs at full load. If the energy cannot be used within the plant, it may even be dumped by

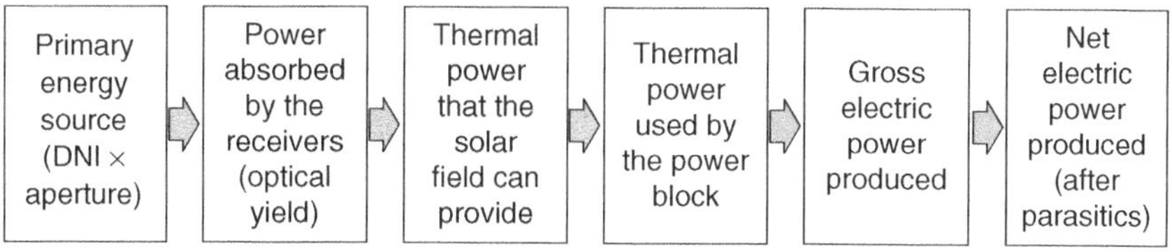

Figure 9.5 Energy conversion steps.

defocusing mirrors, e.g. when storage is full and the power block already runs at full load. Then, the potential solar output is reduced. In principle, cofiring using fossil or biogenic fuels is possible. In the calculations presented in this work, hybrid co-firing is not considered. The power block receives a thermal energy flow from the solar field and/or storage and converts it into electric energy at a certain efficiency depending on the power block operation conditions. Storage is discharged when the solar field cannot supply nominal steam mass flow to the power block. By subtraction of the auxiliary power consumption needed to operate the components, the net power production is calculated to obtain annual values of energy. Annual values are usually more suitable measures for design comparisons (Figure 9.5).

A special focus has been put on assessing effects of power block variables on the system. For this, each design was calculated thermodynamically. Beyond this, for each design, an off-design look-up table was calculated thermodynamically to assess the effect of thermal load, ambient temperature and live steam temperature on performance. Designs with high moisture content at the turbine outlet (endangering the turbine through droplet erosion) are given a high LCOE value in order to sort them out for the optimization process.

A method for air-cooled condenser operation that maximizes the net energy yield for each operation point was developed and applied. This is necessary to evaluate large condensers adequately because they realize favorably low condenser pressures but have, on the other hand, relatively high auxiliary power consumption: Especially at low load cases and low temperatures, cooling cell units need to be turned off to maximize net plant yield (instead of gross plant yield).

9.8.2 Economic Plant Model

Beyond the energetic model of a power plant as described above, the plant is also modeled economically because in many cases, performance and cost are inversely related. By calculating the LCOE, a techno-economic figure of merit is available that considers the effect of any (modeled) technical modification on cost efficiency of the plant (Table 9.1).

Table 9.1 Main technical assumptions.

General plant characteristics	Value/Magnitude
Site/DNI	Daggett, CA (USA) 2791 kWh/(m^2a)
Nominal gross electrical output	55.0 MW
Nominal gross efficiency of the power block	Depending on process parameters, 37.6 $MW_{el,gross}/MW_{th}$
Solar field properties	
Collector type	*Skal-ET*
Aperture width per collector	5.77 m
Optical efficiency relative to DNI on mirror aperture	75.0%
Average cleanliness/field availability	0.97%/99%
HTF temperature at inlet	Depending on live steam pressure, pinch point, and HTF outlet temperature
Receiver thermal losses	New Schott-PTR70 [15]
HTF pressure drop in solar field	Depending on solar field size and mass flow, according to Arias et al. [16]
Specific thermal losses of field piping	Depending on SF temperature (10 W/m^2 at 340°C)
Auxiliary power consumption	
Auxiliary power consumption of solar field	Depending on HTF pressure drop and HTF mass flow (resp. HTF temperature rise)
Auxiliary power of storage	0.003 MW_{el}/MW_{th}
Auxiliary power consumption of power block	Depending on power block design (optimized ACC part load operation)

9.9 Conclusion

The optimization of CSP plant designs through integrated techno-economic modeling presents a significant opportunity to enhance the efficiency, performance, and economic viability of solar energy production. By leveraging advanced modeling techniques that consider both technical and economic factors, CSP plants can be designed to maximize energy output, reduce operational costs, and improve overall sustainability.

Through integrated models, critical design parameters such as solar field size, storage capacity, and plant configuration can be optimized to align with local environmental conditions and economic constraints. This ensures that CSP plants are not only technically feasible but also economically competitive with conventional power generation technologies. Additionally, the use of such models helps in identifying bottlenecks and trade-offs in design choices, enabling more informed decisions.

In conclusion, techno-economic modeling is a powerful tool for CSP plant design optimization, supporting the growth of renewable energy infrastructure. It facilitates the development of cost-effective and high-performance CSP systems, making them a key contributor to global energy transitions and helping to achieve sustainability goals in the energy sector.

References

1 Khalid, M. (2024). Smart grids and renewable energy systems: perspectives and grid integration challenges. *Energy Strategy Reviews* 51: 101299.

2 Soni, N. and Reddy, K.S. (2024). Techno-economic assessment of concentrated solar power technologies integrated with thermal energy storage system for green hydrogen production. *International Journal of Hydrogen Energy* 72: 1184–1203.

3 Li, M., Liu, Y., Li, J. et al. (2024). Optimizing sustainable energy integration: a novel approach using concentrated solar plant and hybrid power supply. *Electric Power Systems Research* 237: 111050.

4 Hossain, M.S., Rahat, M.A.I., Khan, M.S.H. et al. (2024). Techno-economic assessment of various concentrating solar power (CSP) technologies for large-scale sustainable power generation in Bangladesh. *Energy Conversion and Management* 321: 119079.

5 Boretti, A. (2024). Optimizing concentrated solar power: high-temperature molten salt thermal energy storage for enhanced efficiency. *Energy Storage* 6 (7): e70059.

6 Redelinghuys, L.G. and McGregor, C. (2024). Multi-objective techno-economic optimisation of a Carnot battery application in a parabolic trough concentrating solar power plant. *Applied Energy* 376: 123873.

7 Al-Quraan, A. and Al-Mhairat, B. (2024). Economic predictive control-based sizing and energy management for grid-connected hybrid renewable energy systems. *Energy* 302: 131795.

8 Atacama Desert. (2009). Picture from: http://www.ipantalla.com/wp-content/uploads/2009/03/desierto-atacama.jpg (accessed 27 April 2010).

9 Relloso, S. and Delgado, E. (2009). Experience with molten salt thermal storage in a commercial parabolic trough plant. Andasol-1 commissioning and operation. *Proceedings of 15th International SolarPACES Symposium*, Berlin (14–18 September).

10 Strelow, M. (2007). Einsatz numerischer Optimierungsverfahren zur Wirkungsgrad Maximierung des Dampfkraft prozesses solarthermischer Kraftwerke. Diploma thesis. University Braunschweig and Fraunhofer ISE.

11 Morin, G. (2012). Optimisation of concentrating solar power (CSP) plant designs through integrated techno-economic modelling, DOI:10.1533/9780857096173.3.495.

12 Li, G., Du, G., Liu, G., and Yan, J. (2024). Study on the dynamic characteristics, control strategies and load variation rates of the concentrated solar power plant. *Applied Energy* 357: 122538.

13 Khan, M.I., Gutiérrez-Alvarez, R., Asfand, F. et al. (2024). The economics of concentrating solar power (CSP): assessing cost competitiveness and deployment potential. *Renewable and Sustainable Energy Reviews* 200: 114551.

14 Solar Millennium. (2009). Solar Millennium AG, Erlangen, personal e-mail communication by Sven Moormann, Head of Corporate Communications, 13 October 2009.

15 Burkholder, F. and Kutscher, C. (2009). Heat loss testing of Schott's 2008 PTR70 parabolic trough receiver. Report NREL/TP-550-45633.

16 Arias, D. A., Gavilán, A., and Russel, M. (2009). Pumping power parasitics in parabolic trough solar fields. *Proceedings of 15th International SolarPACES Symposium*, Berlin (14–18 September 2009).

Index

d

Printed and bound by CPI Group (UK) Ltd, Croydon, CR0 4YY
17/09/2026
14964332-0001